T0281179

Cambridge Elements

Elements in the Philosophy of Science
edited by
Jacob Stegenga
University of Cambridge

PHILOSOPHY OF PSYCHIATRY

Jonathan Y. Tsou
Iowa State University

CAMBRIDGE
UNIVERSITY PRESS ·

CAMBRIDGE
UNIVERSITY PRESS

University Printing House, Cambridge CB2 8BS, United Kingdom

One Liberty Plaza, 20th Floor, New York, NY 10006, USA

477 Williamstown Road, Port Melbourne, VIC 3207, Australia

314–321, 3rd Floor, Plot 3, Splendor Forum, Jasola District Centre, New Delhi – 110025, India

103 Penang Road, #05–06/07, Visioncrest Commercial, Singapore 238467

Cambridge University Press is part of the University of Cambridge.

It furthers the University's mission by disseminating knowledge in the pursuit of education, learning, and research at the highest international levels of excellence.

www.cambridge.org
Information on this title: www.cambridge.org/9781108706667
DOI: 10.1017/9781108588485

First published 2021

A catalogue record for this publication is available from the British Library.

ISBN 978-1-108-70666-7 Paperback
ISSN 2517-7273 (online)
ISSN 2517-7265 (print)

Philosophy of Psychiatry

Elements in the Philosophy of Science

DOI: 10.1017/9781108588485
First published online: June 2021

Jonathan Y. Tsou
Iowa State University

Author for correspondence: Jonathan Y. Tsou, jtsou@iastate.edu

Abstract: Jonathan Y. Tsou examines and defends positions on central issues in the philosophy of psychiatry. The positions defended assume a naturalistic and realist perspective and are framed against skeptical perspectives on biological psychiatry. Issues addressed include the reality of mental disorders, mechanistic and disease explanations of abnormal behavior, definitions of mental disorder, natural and artificial kinds in psychiatry, biological essentialism and the projectability of psychiatric categories, looping effects and the stability of mental disorders, psychiatric classification, and the validity of the DSM's diagnostic categories. The main argument Tsou defends is that genuine mental disorders are biological kinds with harmful effects. This argument opposes the dogma that mental disorders are necessarily *diseases* (or pathological conditions) that result from biological dysfunction. Tsou contends that the broader ideal of biological kinds offers a more promising and empirically ascertainable naturalistic standard for assessing the reality of mental disorders and the validity of psychiatric categories.

Keywords: natural kinds in psychiatry, the projectability of psychiatric categories, psychiatric classification and diagnostic validity, the nature and reality of mental disorders, disease explanations of abnormal behavior

ISBNs: 9781108706667 (PB), 9781108588485 (OC)
ISSNs: 2517-7273 (online), 2517-7265 (print)

Contents

1 Introduction

This Element examines and defends positions on core issues in philosophy of psychiatry from the perspective of analytic philosophy of science.[1] The positions defended assume a naturalistic and realist perspective and are framed against skeptical perspectives (e.g., anti-psychiatry) that contend that mental disorders are determined by social causes rather than biological ones. Philosophical issues addressed include the reality of mental disorders, the problem of defining mental disorder, disease explanations, natural kinds in psychiatry, feedback effects of psychiatric classifications, the projectability of psychiatric classifications, and the validity of psychiatric constructs.[2]

The central argument of this Element is that mental disorders are biological kinds with harmful effects. This argument implies that genuine mental disorders are natural kinds (i.e., real classes that are discovered by classifiers), rather than artificial kinds (i.e., socially constructed classes that are invented and reflect the values of classifiers). In terms of debates between naturalists and normativists in the philosophy of medicine (Reiss & Ankeny, 2016; Murphy, 2020a), I defend a hybrid account of mental disorder that incorporates naturalistic and normative considerations. Against normativists (e.g., Szasz) who argue that mental disorders are (nothing but) evaluative labels for socially deviant behavior, some mental disorders are identifiable by naturalistic criteria (i.e., biological mechanisms). Against naturalists (e.g., Boorse) who maintain that mental disorders can be identified exclusively by factual criteria, the determination of mental disorders (and physical diseases) requires normative judgments (i.e., that a condition is sufficiently harmful to merit intervention).

The argument that mental disorders are biological kinds opposes the orthodox view that genuine mental disorders are diseases (or pathologies) that result from dysfunctional processes. Against authors (e.g., Wakefield) who define mental disorder in terms of biological dysfunction, I argue that this naturalistic requirement is too restrictive and that the requirement of biological kinds offers a more promising naturalistic standard. This argument also indicates problems with the most influential psychiatric classification manual, the *Diagnostic and Statistical Manual of Mental Disorders* (DSM), whose standardized definitions of mental disorders are utilized worldwide to diagnose individuals in treatment

---

[1] Since the mid-2000s, interest in psychiatry among philosophers of science has grown exponentially. While there were analyses of psychiatry in philosophy of science prior to this period (e.g., see Hempel, 1965, ch. 6; Meehl, 1973, 1991; Hacking, 1995a, 1999, ch. 4), Rachel Cooper's *Classifying Madness* (Cooper, 2005) and Dominic Murphy's *Psychiatry in the Scientific Image* (Murphy, 2006) were agenda-setting works that articulated canonical issues.

[2] Issues that are not directly addressed include, e.g., psychiatric explanation, reductionism, and the role of values in psychiatry. For further issues and readings, see Fulford et al. (2013), Tekin and Bluhm (2019), and Murphy (2020b).

contexts and to compile populations of subjects to study in research contexts. Since 1980, the DSM has implicitly assumed that its objects of classification are disease-syndromes and adopted a purely descriptive approach for defining mental disorders. These features of the DSM have contributed to its failure to formulate *valid* diagnostic categories: definitions that accurately represent ('real') natural classes. I argue that the DSM should classify biological kinds (rather than diseases) and adopt a causal approach to psychiatric classification, which would provide a more transparent and testable system.

In defending a naturalistic approach to philosophy of psychiatry, I engage closely with the scientific literature on mental disorders (e.g., genetics, neuroscience, pharmacology, psychology, anthropology). For numerous reasons, my analysis focuses on schizophrenia and depression as examples. First, schizophrenia and depression are paradigm cases of mental disorders. If any conditions studied in psychiatry are genuine mental disorders, schizophrenia and depression are. Szasz (1988) argues that schizophrenia is psychiatry's "sacred symbol" insofar as it is employed to justify psychiatry's general theories and treatment models. Similarly, Lilienfeld and Marino (1995) contend that mental disorder is a family-resemblance concept organized around ideal prototypes, such as schizophrenia and depression. Second, scientific research on the genetic and neurobiological underpinnings of schizophrenia and depression is much more developed and mature when compared to the research performed on other mental disorders.

Two philosophical accounts that inform my analysis are mechanisms and homeostatic property cluster (HPC) kinds. My discussion of biological mechanisms involved in mental disorders follows the "new mechanical philosophy" (Glennan, 2017; Craver & Tabery, 2019), which advocates analyzing scientific phenomena in terms of mechanisms. In this tradition, mechanisms are understood as complex systems made up of interacting parts that are organized such that they produce regular changes in a phenomenon (e.g., see Machamer, Darden, & Craver, 2000; Glennan, 2002; Bechtel & Abrahamsen, 2005; Bechtel & Richardson, 2010). The concept of mechanisms is motivated to replace older philosophical concepts (e.g., explanation, reduction, laws of nature, causality) with mechanistic accounts that can better account for the concepts and practices adopted in the sciences, especially the biological and psychological sciences (cf. Ross, 2021).

My discussion of natural kinds in psychiatry (and the reality of mental disorders more generally) draws on the account of HPC kinds defended by Boyd (1999a). I argue that the distinction between natural kinds and artificial kinds is a distinction of degree, rather than kind. On this view, *more natural* kinds of mental disorders (e.g., schizophrenia, bipolar disorder) are biological kinds primarily underwritten by biological mechanisms. By contrast, *more artificial* kinds (e.g., hysteria, voyeuristic disorder) are social kinds primarily underwritten by social mechanisms. Assuming that the naturalness (or 'reality') of mental disorders stems from their biological basis, my account is more restrictive than Boyd's insofar as it demands that at least *some* of the mechanisms underwriting mental disorders are intrinsic (biological) mechanisms. This implies that natural kinds in psychiatry are constituted by a *partly intrinsic biological essence* (Devitt, 2008).

In section 2, I critically examine important skeptical perspectives on psychiatry that emerged in the 1960s (i.e., anti-psychiatry, labeling theory). Skeptics argue that mental disorders are not diseases or biological kinds, and they suggest that mental disorders are best explained in terms of social factors. For example, Szasz and Laing contend that 'schizophrenia' is not a disease, but an oppressive label used to explain socially deviant behavior. I counter these claims by appealing to scientific evidence indicating that schizophrenia and depression are underwritten by identifiable biological mechanisms.

In section 3, I examine the prominent definitions of mental disorder defended by Boorse and Wakefield, which aim to delineate the proper objects of study in psychiatry. Boorse's naturalist ("biostatistical") theory of disease implies that genuine mental disorders are conditions that impair normal mental functioning. Wakefield's hybrid ("harmful dysfunction") account maintains that genuine mental disorders involve the failure of a mental mechanism to perform its naturally selected function, which results in harm to individuals. Boorse's account of biological dysfunction is superior to Wakefield's insofar as it avoids the uncritical ('adaptationist') assumptions renounced by philosophers of biology (e.g., Gould). Wakefield's hybrid account is preferable to Boorse's in its requirement that mental disorders must satisfy a normative harm condition. Against both of these authors, I argue—on pragmatic grounds—that the naturalistic requirement of biological dysfunction (which implies mental disorders are *diseases*) is too demanding. A more useful and empirically ascertainable naturalistic requirement is that mental disorders should be biological kinds. Hence, mental disorders are biological kinds whose effects lead to harmful consequences.

In section 4, I argue that some mental disorders (e.g., schizophrenia, depression) are HPC kinds: classes of abnormal behavior constituted by a set of stable biological mechanisms. The naturalistic requirement that HPC kinds are underwritten by intrinsic biological mechanisms ensures that HPC kinds are relatively stable objects of classification and their classifications yield *projectable inferences* (i.e., reliable predictions about kind members). This perspective indicates limitations of Hacking's argument that the objects of classification in psychiatry are inherently unstable because of the 'looping effects' of human kinds. I subsequently examine how biological mechanisms interact with the social mechanisms highlighted by Hacking and skeptics of psychiatry. I argue that biological mechanisms determine the general structural features of mental disorders (e.g., psychosis, depressive states), whereas social mechanisms determine the specific expression of such disorders. This framework illuminates cross-cultural findings, historically bound ('transient') mental disorders, and culture-bound syndromes.

In section 5, I examine issues regarding the failure of the DSM to formulate valid diagnostic categories that correspond to real classes in nature. After reviewing several concepts of diagnostic validity, I argue that only a few DSM categories (e.g., 'schizophrenia,' 'bipolar disorder') possess *construct validity*—that is, they accurately represent a well-confirmed theoretical construct. The main reason for this failure is the DSM's theoretical assumption that mental disorders are *discrete disease-syndromes caused by dysfunctional processes*. Accordingly, I contend that the DSM should classify biological kinds (i.e., HPC kinds), rather than diseases. I subsequently examine a philosophical criticism of the DSM articulated by Tabb, which contrasts the methodological assumptions of the DSM with those of the *Research Domain Criteria* (RDoC). Tabb argues that the DSM has failed (and is unlikely) to formulate valid diagnostic categories because of the way that DSM definitions are utilized to assemble groups of patients to study: DSM definitions produce *heterogenous* groups that are unsuitable for the discovery of biomedical facts. Tabb's argument neglects a more fundamental problem with the DSM: its failure to revise its diagnostic categories to incorporate scientific findings. More generally, the DSM should adopt a theoretical and etiological (as opposed to purely descriptive) approach to classification, which would provide a more promising method for formulating valid psychiatric categories.

Given the centrality of some controversial arguments defended in this Element, I bring attention to them here. A dogma of contemporary psychiatry is that mental disorders are a subset of medical diseases.[3] I reject this assumption in my argument that genuine mental disorders are biological kinds with harmful effects (sections 3.5, 4.5, and 5.3). While the stipulation that mental disorders are biological kinds includes diseases as a subclass, I propose that mental disorder is a broader category that includes conditions (e.g., psychological reactions)

---

[3] This assumption stems from the medical model that understands abnormal behaviors as *symptoms of disease* (Guze, 1992). While some mental health professionals (e.g., cognitive-behavioral therapists, humanistic psychologists, social workers) reject this assumption, it is deeply ingrained in the formal language (e.g., "mental health," "psychopathology," "treatment") and institutions (e.g., the National Institute of Mental Health) surrounding the study of abnormal psychology.

caused by biological mechanisms falling within the normal range of biological variation, rather than dysfunctional mechanisms per se.

My rejection of the assumption that genuine mental disorders are diseases is related to arguments that I make about depression. I suggest that some forms of depression (e.g., acute depression, bereavement) are normal biological and psychological responses, rather than conditions caused by dysfunctional mechanisms. These types of depression can be understood as *normal ('universal') psychological reactions to stress and trauma*, rather than diseases (cf. Lilienfeld & Marino, 1995; Wakefield, 1999). While some forms of depression are caused by dysfunctional processes (e.g., chronic and persistent forms of depression), the DSM category of major depressive disorder is insensitive to such distinctions (Horwitz & Wakefield, 2007).[4]

Another controversial argument concerns the very distinction between depression and anxiety. I criticize the DSM's operational definition of depression for its exclusion of anxiety as a characteristic sign. Research indicates that anxiety is a core sign of depression and that depression and anxiety co-occur very frequently (sections 2.2 and 5.2). From a classificatory standpoint, this would motivate collapsing the DSM categories of mood and anxiety disorders into a single category, perhaps distinguishing depression-anxiety that tends toward distress versus fear (see note 11). While I discuss depression as distinct from anxiety for purposes of clarity, this qualification should be kept in mind.

2 Skepticism about Biological Psychiatry

In the 1960s, anti-psychiatrists and labeling theorists advanced influential arguments that mental disorders are better explained by social factors than biological ones. For example, Szasz and Laing maintain that 'schizophrenia' does not refer to a disease, but a strategy adopted by individuals to cope with 'problems in living.' Against the more radical conclusions reached by these authors and the argument that social mechanisms should be privileged over biological mechanisms in explanations of mental disorder, I examine scientific evidence supporting the presence of identifiable biological mechanisms underwriting mental disorders such as schizophrenia and depression. While social accounts of mental disorder should not supplant biological accounts, the social

---

[4] Although the DSM distinguishes more chronic forms of depression—what has variously been called "dysthymia," "depressive personality disorder," or "persistent depressive disorder" (Hirschfeld, 1994; Rhebergen & Graham, 2014)—it currently conceptualizes *all* forms of depression as diseases. Curiously, the first two editions of the DSM, which were informed by psychoanalysis, drew an etiological distinction between *biological diseases* and *psychological reactions* (Tsou, 2011, 2019).

mechanisms emphasized in these accounts (e.g., role adoption) play a role in in the presentation of the signs of mental disorders.

2.1 Anti-Psychiatry and Labeling Theory

Thomas Szasz is a psychoanalytic psychiatrist who is regarded as the main figurehead of the anti-psychiatry movement.[5] His argument that mental illness is a 'myth' maintains that the explanatory concept of mental illness is invalid. In his seminal paper, Szasz (1960) rejects 'mental illness' as an explanatory concept that is posited to be the cause of various abnormal behaviors (e.g., 'schizophrenia,' 'depression'):

> My aim in this essay is to raise the question "Is there such a thing as mental illness?" and to argue that there is not ... Mental illness ... is not literally a "thing"—or physical object—and hence it can "exist" only in the same sort of way in which other theoretical concepts exist.... During certain historical periods, explanatory conceptions such as deities, witches, and microorganisms appeared not only as theories but as self-evident *causes* of a vast number of events. I submit that today mental illness is widely regarded in a similar fashion, that is, as the cause of innumerable diverse happenings. (p. 113)

In rejecting this explanatory concept, Szasz complains that it offers a simplistic and convenient explanation that obscures the genuine causes of abnormal behavior, viz., conflicts in human relations or 'problems in living.'

In other works, Szasz presents his argument as the claim that mental illness is a "metaphorical disease," as opposed to a real or factual disease (Szasz, 1974, pp. x–xi). This position is stated in the first proposition of Szasz's "manifesto" (Szasz, 1998):

> Mental illness is a metaphor (metaphorical disease). The word "disease" denotes a demonstrable biological process that affects the bodies of living organisms (plants, animals, and humans). The term "mental illness" refers to the undesirable thoughts, feelings, and behaviors of persons. Classifying thoughts, feelings, and behaviors as diseases is a logical and semantic error.

Szasz distinguishes between physical illnesses (which are objectively identified) and mental illnesses (which are subjectively identified). Unlike genuine (physical) illness, which involves identifiable biological processes, mental illness merely involves disapproval. In this framework, "mental illness" refers to thoughts, feelings, and behaviors *that are disapproved of by society*. For Szasz (1960), this marks a significant difference in the *norms* appealed to in

---

[5] Anti-psychiatry is neither a homogenous nor well-defined movement. Szasz (2004) attempted to distance himself from the anti-psychiatry label and the views of R. D. Laing in particular.

determining physical illness (objective biological criteria) and mental illness (subjective social criteria):

> The concept of illness, whether bodily or mental, implies *deviation from some* clearly *defined norm*. In the case of physical illness, the norm is the structural and functional integrity of the human body.... [W]hat health *is* can be stated in anatomical and physiological terms. What is the norm deviation [involved in] mental illness? ... [W]hatever this norm might be, we can be certain ... that it is a norm that must be stated in terms of *psycho-social, ethical,* and *legal* concepts. (p. 114)

According Szasz, the identification of mental illness (deviation from social and ethical norms) is subjective, whereas the identification of physical illness (deviation from biological norms) is objective and intersubjective. This fundamental difference renders it illegitimate to subsume mental illness under the more general category of (physical) illness.

Against the assumption that some classes of abnormal behavior are indicative of disease processes, Szasz argues that these classes are indicative of *problems in living*. Szasz (1960) writes:

> [I]n our scientific theories of [abnormal] behavior—we have failed to *accept* the simple fact that human relations are inherently fraught with difficulties.... [T]he idea of mental illness is now being put to work to obscure certain difficulties which ... may be inherent ... in the social intercourse of persons.... [I]nstead of calling attention to human needs, aspirations, and values, the notion of mental illness provides an amoral and impersonal "thing" (an "illness") as an explanation for *problems in living*. (p. 117)

Problems in living refer to the burden of choosing how to conduct our lives. Szasz (1974) articulates a positive account of abnormal behavior (a 'theory of personal conduct') within a game framework (Mead, 1934), wherein approved of conduct is understood as conducting one's life in a personally and socially satisfying manner. In this framework, role-taking behavior is a ubiquitous aspect of human life. However, individuals may adopt better or worse roles (i.e., play games well or poorly). Szasz (1974) explains abnormal behavior as *impersonation*, viz., inconsistent or dishonest role-playing:

> So-called mental illnesses are best conceptualized as special instances of impersonation. In hysteria, for example, the patient impersonates the role of a person sick with the particular disease or disability he displays ... The hysteric's seeming ignorance of what he is doing may ... be interpreted as his not being able to afford to know it ... He must therefore lie both to himself and others.... The so-called hypochondriac and schizophrenic also impersonate: the former takes the role of certain medical patients, whereas the latter

often takes the role of other, invariably famous, personalities. . . . Roles can and do become habits. In many chronic cases of mental illness, we witness the consequences of playing hysterical, hypochondriacal, schizophrenic, or other games . . . until they have become deeply ingrained habits. (pp. 237–9)

Szasz argues that abnormal behavior is a special strategy (i.e., 'impersonation') that individuals adopt as a last resort to gain help. In this sense, Szasz understands abnormal behaviors to be caused by problems in living: "[P]ersons called mentally ill . . . impersonate the roles of helplessness, hopelessness, weakness, and often bodily illness—when, in fact, their actual roles pertain to frustrations, unhappiness, and perplexities due to interpersonal, social, and ethical conflicts" (Szasz, 1974, p. 246). Hence, abnormal behavior should be explained and understood as a kind of role adoption that is caused by problems in living, not disease processes.

R. D. Laing is another anti-psychiatrist, trained in psychoanalysis, who rejects disease explanations of abnormal behavior in favor of social and psychological explanations. Being influenced by existential philosophy, Laing's analyses focus on the phenomenology or experience of abnormality. In analyzing the experience of schizophrenia, Laing emphasizes the social context surrounding mental patients. Citing Erving Goffman (1961) approvingly, Laing (1967) writes: "[Goffman] does not just describe [abnormal] behavior 'in' mental . . . patients, he describes it within the context of personal interaction and the system in which it takes place" (p. 111). Laing also argues that the 'double-bind hypothesis' identifies an important cause of schizophrenia: early in life, schizophrenics are subjected to double-bind communications that convey mixed or contradicting messages (Bateson et al. 1956).[6] Laing (1967) suggests that this 'can't win' situation leads to schizophrenia: "[*W*]*ithout exception* the experience and behavior that gets labeled schizophrenic is a special strategy that a person invents in order to live in an unlivable situation . . . the person has come to feel he is in an untenable position" (pp. 114–5). Like Szasz, Laing presents schizophrenia as a strategy that a person adopts—consciously or unconsciously—in response to an existential crisis. In championing this account of schizophrenia as a coping strategy, Laing (1967) asserts that:

[T]here is no such "condition" as "schizophrenia" but the label is a social fact [which itself is a] political event [that] . . . imposes definitions and consequences on the labeled person. It is a social prescription that rationalizes a set of social actions whereby the labeled person is annexed by others . . . The

---

[6] The double-bind hypothesis of schizophrenia has been heavily criticized and attempts to corroborate it have been unsuccessful (Schuham, 1967).

person labeled is inaugurated not only into a role, but into the career of a patient. (Laing, 1967, p. 121–2)

Like Szasz, Laing rejects disease explanations of schizophrenia. His favored social explanation implies that labeling a person as 'schizophrenic' ostracizes them further and exacerbates difficulties in their lives.

Thomas Scheff is a sociologist of deviance who explains the development of mental illness in terms of social mechanisms. Like Szasz and Laing, Scheff presents 'mental illness' as a special *role* that individuals adopt in response to social pressures (Scheff, 1963). His main question is: what are the conditions under which diverse kinds of deviance become stable and uniform? Scheff addresses this question with reference to the category of *residually deviant behavior* (RDB): deviant behaviors that do not fall under established categories (e.g., criminality, perversion, bad manners). Scheff presents the symptoms of mental illness as RDB, and he aims to explain why residual deviance becomes stable and uniform for some individuals. Scheff maintains that the origins of residual deviance are diverse, including biological causes, psychological differences, external stresses, and acts of volition.[7] Regardless of its origins, most RDB is temporary and transitory, and it is often unrecognized, denied, or rationalized (e.g., as eccentricity) by individuals. This leads Scheff (1963) to his main concern:

> If residual deviance is highly prevalent among ostensibly "normal" persons and is usually transitory . . . what accounts for the small percentage of residual deviants who go on to deviant careers? . . . The conventional hypothesis is that that the answer lies in the deviant himself. The hypothesis suggested here is that the most important single factor . . . is the societal reaction. Residual deviance may be stabilized if it is defined to be evidence of mental illness, and/or the deviant is placed in a deviant status, and begins to play the role of the mentally ill. (p. 442)

Hence, Scheff contends that the most important factors responsible for stabilizing residual deviance are social: labeling and role adoption.

Scheff appeals to labeling theory to explain the stabilization of RBD. He argues that traditional stereotypes about mental illness are learned early in life and constantly reinforced by society. During a crisis situation, these stereotypes function to direct an individual's behavior:

> [T]he traditional stereotypes of mental disorder are solidly entrenched . . . because they are learned early in childhood and are continuously reaffirmed

---

[7] Scheff (1963) writes: "It appears likely . . . that there are genetic, biochemical, or physiological origins for residual deviance" (p. 439). Laing (1967) similarly acknowledges a circumstantial role for biochemistry in schizophrenia (p. 115).

in the mass media and in everyday conversation … In a crisis, when the deviance of an individual becomes a public issue, the traditional stereotype of insanity becomes the guiding imagery for action, both for those reacting to the deviant and, at times, for the deviant himself. When societal agents and persons around the deviant react to him uniformly in terms of the traditional stereotypes of insanity, his amorphous and unstructured deviant behavior tends to crystallize in conformity to these expectations … The process of becoming uniform and stable is completed when the traditional imagery becomes a part of the deviant's orientation for guiding his own behavior. (Scheff, 1963, pp. 447–8)

Scheff (1963) contends that "most mental disorder can be considered to be a social role" (p. 449). The role of being mentally ill is stabilized when an individual *accepts* the label of mental illness.[8] In discussing the causal processes that lead individuals to accept such a role, Scheff notes that labeled deviants are often rewarded for playing a stereotyped role (e.g., receiving praise for agreeing with a psychiatrist) and punished for attempting to return to conventional roles (e.g., facing discrimination when trying to reenter the workplace). Moreover, during a crisis stage when the deviant is labeled as 'mentally ill,' individuals are highly suggestible (especially to social cues) and may accept the mentally ill role as the only option. Scheff (1963) concludes that: "Among residual deviants, labeling is the single most important cause of careers of residual deviance" (p. 451).

2.2 A Naturalistic Response to Skeptics of Biological Psychiatry

Skeptics of biological psychiatry argue that classes of abnormal behavior (e.g., schizophrenia, depression) are better explained by social and psychological causes (e.g., labeling, role-adoption) than biological ones. I articulate a naturalistic response to these skeptics by challenging the claims that: (1) mental illness is an invalid explanatory concept, and (2) abnormal behavior is better explained by social causes than biological ones. Against (2), I argue that the signs of schizophrenia and depression are underwritten by demonstrable biological processes.

Szasz's argument that mental illness is a myth is unconvincing because it presupposes an uncharitable (stipulative) definition of 'mental illness.' This is evident in his manifesto, where Szasz (1998) distinguishes:

(1) bodily illness: a demonstrable biological process that affects the bodies of persons.

---

[8] In section 4.2, I discuss Hacking's philosophical argument that emphasizes similar social mechanisms. Ron Mallon (2016) articulates a naturalistic account of how 'social roles' become entrenched that complements these analyses.

(2) mental illness: the undesirable thoughts, feelings, and behaviors of persons.

This formulation equivocates on the meaning of 'illness.' In (1), 'illness' refers to the *cause of symptoms*, whereas 'illness' refers to the *symptoms of illness* in (2). A more consistent and charitable formulation of (2) would be:

(2′) mental illness: a demonstrable biological process that affects the thoughts, feelings, and behaviors of persons.[9]

Szasz fails to establish that (2′) is an invalid explanatory concept. Against Szasz's denial that schizophrenia and depression are explainable by demonstrable biological processes (e.g., see Szasz, 1988, 1997, ch. 3), I present scientific evidence that they are.

Contemporary skeptics of psychiatry—who argue from the perspectives of 'postpsychiatry' (Bracken & Thomas, 2005) and 'critical psychiatry' (Middleton & Moncrieff, 2019)[10]—endorse Szasz's argument that mental illness (2′) is an invalid explanatory concept. In a programmatic paper, authors allied with these movements assert that biological psychiatry is founded on the false assumptions that (Bracken et al. 2012, p. 430):

(a) Mental health problems arise from mechanisms or processes involving abnormal physiological or psychological events occurring within the individual.
(b) These mechanisms or processes can be modeled in causal terms. They are not context-dependent.

While contemporary skeptics are correct that the mechanisms in (b) are 'context-dependent,' they are incorrect to assume that they cannot be causally modeled (or that mechanisms in medicine are any different). For schizophrenia

---

[9] Szasz (1960) would object that (2′) refers to brain disease (i.e., bodily illness). Naturalists who reject dualism, however, should not feel compelled to follow Szasz's stipulation that mental illnesses must have mental causes (cf. Cooper, 2007, ch. 2). As a materialist, I maintain that the 'mental' in 'mental illness' (or 'mental disorder') should refer to abnormal *psychological* ('mental') *symptoms* (or signs), e.g., psychosis, mania (cf. Murphy, 2006, ch. 3; Radden, 2019). Szasz (1960) would also object that the 'mental symptoms' in (2′) cannot be objectively determined because they refer to individuals' *communications* (p. 114). I assume that individuals' self-reports (e.g., of psychosis or mania) are a form of evidence that can be corroborated and identified objectively. Moreover, Szasz's argument that physical symptoms are identified objectively, whereas mental symptoms are identified subjectively because they appeal to social norms, rests on an unrealistic (value-free) view of bodily illness. Arguably (see section 3), *all* attributions of illness (mental or physical) require value-judgments regarding the *undesirability or harmfulness* of a condition based on social norms (Sedgwick, 1973; Wakefield, 1992b).

[10] Critical psychiatry is the broader field, while postpsychiatry is a narrower field more explicitly allied with postmodernism. Critical psychiatry has usefully moved beyond anti-psychiatry by focusing on showing that particular mental disorders (e.g., depression, schizophrenia, bipolar disorder) are not diseases, highlighting the limited efficacy and unrecognized harms of psychiatric drug interventions, and advocating collaboration with the service user movement (Bracken et al., 2012).

and depression, there is robust evidence for the presence of demonstrable biological mechanisms that cause abnormal psychological processes and behaviors (e.g., psychosis, depressive states). This demonstrates that the explanatory concept of mental illness (2′) rejected by Szasz and his enthusiasts is valid for some mental disorders.

Schizophrenia is a relatively rare mental disorder characterized by disruptions of thought, feeling, and behavior. Schizophrenia is defined in DSM-5 by the following symptoms (APA, 2013, pp. 99–100):

(1) delusions
(2) hallucinations
(3) disorganized speech
(4) grossly disorganized behavior or catatonic behavior
(5) diminished emotional expression or lack of motivation

(1) and (2) are positive (or psychotic) symptoms; (3) and (4) are disorganized symptoms; and (5) are negative symptoms (Kring et al., 2017, pp. 252–8). Positive symptoms (i.e., psychosis) are present in schizophrenic episodes and regarded as the classic signs of schizophrenia (cf. Lewis, 2012). Delusions are typically organized around a coherent theme (e.g., paranoid, persecutory); hallucinations are typically auditory. Whereas positive (and disorganized) symptoms are psychological excesses, negative symptoms are psychological deficits. The prevalence of schizophrenia worldwide is estimated to be less than 1% of the overall population, with prevalence estimates ranging between 0.4–0.7% (Saha et al. 2005). Lifetime prevalence rates of schizophrenia (i.e., the proportion of people who have ever experienced schizophrenia) are estimated to be 0.43% (McGrath et al., 2008). Anthropological research indicates that the core symptoms of schizophrenia (i.e., delusions, hallucinations, association of unrelated ideas, social withdrawal, lack of motivation, and lack of emotional response) appear in all cultures (Kleinman, 1988, ch. 3).

Since the 1960s, evidence has accumulated that schizophrenia is underwritten by stable neurobiological mechanisms. Research on dopamine pathways, led to the articulation and revision of the 'dopamine hypothesis' (Howes & Kapur, 2009). This theory maintains that positive and disorganized symptoms (e.g., psychosis, disorganized thought) are caused by excessive dopamine activity in the mesolimbic pathway, which originates in the ventral tegmental area (VTA) and projects to subcortical areas such as the nucleus accumbens, hypothalamus, amygdala, hippocampus, and striatum. By contrast, negative symptoms (e.g., avolition, flat affect) are caused by deficient dopamine activity in the mesocortical pathway, which originates in the VTA and projects to cortical areas such as the prefrontal cortex (Kring et al., 2017, pp. 264–6;

Hancock & McKim, 2018, pp. 76–7). Multiple pharmacological findings support this hypothesis (Tsou, 2012, 2017). Evidence that positive symptoms are caused by excessive dopamine activity in the mesolimbic pathway is supported by the fact that antipsychotic drugs that can treat positive symptoms (i.e., psychosis) are dopamine antagonists that decrease dopamine activity in this pathway through their blockade effect on D_2 dopamine receptors. Moreover, dopamine agonist drugs (i.e., stimulants) with opposite pharmacological effects of antipsychotic drugs, when taken in sufficiently large doses, induce an 'amphetamine psychosis,' which is indistinguishable from positive symptoms and treatable with antipsychotics. Evidence that negative symptoms result from deficient dopamine activity in the mesocortical pathway derives from research focused on the prefrontal cortex (PFC), the terminal region of the mesocortical pathway. This research identifies robust correlations between underactive dopamine neurons in the PFC and the negative symptoms (and cognitive impairments) observed in schizophrenia (Kring et al., 2017, pp. 264–8). Deficient dopamine activity in the PFC is theorized to disinhibit control over dopamine activity in subcortical areas (e.g., the amygdala), resulting in excessive dopamine activity in the mesolimbic pathway (Weinberger, 1987).

Research implicates other neurotransmitters (e.g., glutamate) and mechanisms (e.g., neuroplasticity) involved in schizophrenia. Dopaminergic systems are modulated by neurotransmitter systems (e.g., glutamate, serotonin, norepinephrine) whose afferents terminate and converge on dopamine neurons in the VTA (Hancock & McKim, 2018, p. 76). Since the 1990s, attention has focused on the 'glutamate hypothesis' that schizophrenia is caused by deficient glutamate activity in the PFC involving *N*-methyl-D-aspartate (NMDA) receptors (Moghaddam & Javitt, 2012). This hypothesis is supported by the finding that glutamate antagonist drugs (e.g., PCP, ketamine), which decrease the activity of glutamate through their blockade of NMDA glutamate receptors, can produce positive, disorganized, and negative symptoms (Javitt, 2007). Glutamate agonist drugs (e.g., glycine, *d*-serine) —when added as an adjunctive medication to an antipsychotic drug—can be effective at improving negative symptoms and cognitive impairments; there is limited evidence that glutamate agonists improve positive symptoms (Javitt, 2007). Some researchers (e.g., Olney & Farber, 1995) argue that deficient glutamate activity is a 'primary deficit' that results in downstream effects on dopamine systems. This etiological interpretation is likely an oversimplification given that dopamine activity—through interactions with GABA systems—modulates glutamate transmission (Guillin, Abi-Dhargham, & Laruelle, 2007). Glutamate and dopamine mechanisms likely interact with one another (and other neurotransmitters) in a complex manner to produce the signs of schizophrenia (Laruelle, Kegeles, & Abi-Dargham, 2003; Kendler & Schaffner, 2011). More recent research on neuroplasticity (i.e., the capacity of neural networks to adapt through growth and reorganization) suggests that impairments in neuroplasticity (e.g., reduced dendritic spine density of neurons in the dorsolateral PFC) are responsible for deficient glutamate activity (Lewis & González-Burgos, 2008).

Genetics research indicates that schizophrenia is a highly heritable disorder, and there is limited evidence about genetic mechanisms. Twin studies produce heritability estimates (i.e., the proportion of variation in a disorder attributable to genetic factors) around 81% (Sullivan, 2005). Molecular genetic studies (e.g., linkage and association studies) provide limited evidence for some candidate genes (e.g., *NRG1*, *DISC1*, *RGS4*), including genes related to neurobiological systems implicated in positive symptoms (e.g., *DTNBP1*) and negative symptoms (*COMT, BDNF*). However, results are inconclusive and non-candidate gene driven research, such as genome-wide association studies (GWAS), have not independently implicated these genes (Kim et al., 2011). GWAS have provided two robust findings: (1) genetic variation in the major histocompatibility complex (MHC) on chromosome 6, and (2) genetic variation near *MIR*137 (a non-coding RNA molecule) on chromosome 1 (The Schizophrenia Psychiatric Genome-Wide Association Study [GWAS] Consortium, 2011). Molecular genetics research supports the inference that schizophrenia is a polygenic disorder constituted by common genetic variants at multiple loci that each have a small effect on risk. There is also evidence of rarer genetic mutations or copy number variations (e.g., a deletion of a section of DNA in a gene) that can lead to larger effects on risk (Kim et al., 2011). GWAS on single-nucleotide polymorphisms or SNPs (i.e., differences in single nucleotides in the DNA sequence of a particular gene) indicate that the SNPs involved in schizophrenia are also implicated in bipolar disorder, suggesting a common genetic vulnerability for both disorders (Sullivan, Daly, & O'Donovan, 2012).

Depression is a relatively common disorder characterized by profound feelings of sadness and an inability to experience pleasure. Major depressive disorder is defined in DSM-5 by the following criteria (APA 2013, pp. 160–1):

1. Depressed mood (sadness, joylessness)
2. Loss of interest or pleasure
3. Significant changes in weight or appetite
4. Sleep disturbances
5. Psychomotor agitation or retardation
6. Fatigue or loss of energy
7. Feelings of worthlessness or guilt
8. Diminished ability to think or concentrate, or indecisiveness
9. Recurrent thoughts of death or suicide

(1) and (2) are regarded as the cardinal signs of depression. Compared to prevalence rates of schizophrenia and bipolar disorder, which are consistently estimated to be less than 1% of the population, prevalence rates of depression are considerably higher (and more variable across cultures and time). In 2015, it was estimated that the depression afflicts 4.4% of the population worldwide, with higher prevalence rates for women than men; moreover, the incidence of depression is estimated to have increased 18.4% worldwide between 2005 to 2015 (WHO, 2017). Lifetime prevalence of depression have been estimated to be around 15% (compared to 1.1% for bipolar disorder), making depression one of the most common mental disorders (Merikangas et al., 2007). Depression is observed in all cultures, although its prevalence and typical expression vary widely across cultures (Kleinman, 1988, ch. 3; Weissman et al., 1996). A WHO cross-cultural study identified a 'common core' of depressive signs found among 75% of depressed patients across the world: sadness, joylessness, anxiety, tension, lack of energy, decreased interest and concentration, feelings of inadequacy, and feelings of worthlessness (Sartorius et al. 1983, p. 92).[11] DSM-5 distinguishes depression from a chronic subtype, persistent depressive disorder (formerly dysthymic disorder); the prevalence of dysthymia (e.g., lifetime prevalence rates estimated at 2.5%) is significantly lower than major depression (Kessler et al., 2005).

Since the 1960s, evidence has accumulated that depression is underwritten by demonstrable neurobiological mechanisms. The 'monoamine hypothesis' maintains that depression is associated with deficient activity of monoamine neurotransmitters: serotonin, norepinephrine, and dopamine. Numerous pharmacological findings support this hypothesis (Tsou, 2013, 2017).[12] First-generation antidepressant drugs—i.e., monoamine oxidase inhibitor and tricyclic antidepressant drugs—which can relieve the signs of depression, are

---

[11] Among the core signs of depression identified in the WHO study, anxiety is noticeably absent from the DSM's diagnostic criteria for major depression. Some research (e.g., epidemiology, genetics) suggests that the DSM's distinction between depression and anxiety is unsupported by empirical evidence. Around 60% of individuals with depression experience an anxiety disorder in their lifetime; around 60% of individuals with an anxiety disorder experience depression in their lifetime (Kessler et al., 2003; Moffitt et al., 2007). Depression is especially likely to co-occur with generalized anxiety disorder (Watson, 2009). Moreover, the genetic risk of depression and generalized anxiety disorder overlap considerably (Kendler et al., 2003). Prior to the publication of DSM-5 (APA, 2013), some researchers (Watson, O'Hara, & Stuart, 2008) argued that DSM-5 should subsume mood disorders and anxiety disorders into a single category and distinguish these disorders based on their propensity towards unhappiness and distress (e.g., depression, generalized anxiety) versus fear (e.g., panic disorder, phobias, bipolar disorder). These recommendations were not adopted.

[12] Stegenga (2018) emphasizes that the efficacy of antidepressants is modest, although he maintains that modest effectiveness is typical of most medical interventions. My analysis only assumes that pharmacological interventions with antidepressants provide (fallible) evidence about the neurobiological mechanisms underlying depression (Tsou, 2017).

monoamine agonist drugs that increase the activity of monoamine neurotrans-mitters. Moreover, monoamine antagonist drugs with opposite pharmacological effects of first-generation antidepressants (e.g., reserpine) can induce depres-sive symptoms (Hancock & McKim, 2018, p. 292). Given the clinical efficacy of second-generation antidepressants—i.e., selective serotonin reuptake inhibi-tors (SSRIs)—which were developed in the 1980s following a procedure of targeted design (Wong et al., 2005), particular attention has focused on serotonin.

The 'serotonin hypothesis' holds that the feelings of sadness and loss of interest associated with depression are related to deficient serotonin activity in the serotonin pathway that originates in the raphe nuclei and projects to the neocortex, thalamus, basal ganglia, hippocampus, and amygdala (Hancock & McKim, 2018, pp. 74–7). This hypothesis is supported by the fact that serotonin agonist drugs can relieve depression. Moreover, tryptophan depletion studies demonstrate that lowering levels of tryptophan (the main precursor of sero-tonin) with a tryptophan-free amino acid drink (which reduces serotonin activ-ity) causes acute depressive symptoms for subjects with a history of depression, but not for subjects without such a history (Hancock & McKim, 2018, p. 293). Depletion studies appear to clarify a trait shared by individuals with a vulner-ability to depression (viz., insensitive 5-HT_{1A} postsynaptic receptors), rather than a state-dependent change caused by serotonin depletion (Kring et al., 2017, pp. 145–6). The serotonin hypothesis is also supported by the finding that decreased serotonin activity is correlated with decreased cerebrospinal fluid (CSF) levels of 5-hydroxyindoleacetic acid (5-HIAA), which is the main metabolite of serotonin. Studies reveal significantly lower CSF levels of 5-HIAA among depressed individuals (Hancock & McKim, 2018, p. 292).

While deficient serotonin activity appears to be involved in depression, deficient dopamine and norepinephrine activity are also likely involved (Nestler & Carlezon, 2006; Moret & Briley, 2011). This view is supported by the efficacy of third-generation antidepressants: serotonin and norepinephrine reuptake inhibitors (SNRIs) developed in the mid-1990s and serotonin-norepinephrine-dopamine reuptake inhibitors (SNDRIs) developed in the 2010s (Hancock & McKim, 2018, pp. 296–300). There is evidence that defi-cient norepinephrine and dopamine activity is related to the lack of energy, fatigue, and lack of motivation associated with depression (Stahl, 2002).

Research implicates specific brain areas and other mechanisms (e.g., neuro-endocrine response, neuroplasticity) involved in depression. Neuroimaging studies indicate that depression is associated with elevated activity in the amygdala and anterior cingulate; and diminished activity in the PFC, hippo-campus, and striatum (Davidson et al., 2002). Neuroendocrinological studies have revealed irregularities in hypothalamic–pituitary–adrenal (HPA) axis, which is the biological system that manages stress (Kring et al., 2017, pp. 147–8). The 'glucocorticoid theory' maintains that hyperactivity of the HPA axis and elevated levels of the stress hormone cortisol (which is involved in the fight-flight response) are related to depression (Hancock & McKim, 2018, pp. 295–6). One robust finding is a correlation between stress, high levels of cortisol, and a reduction of function of postsynaptic 5-HT$_{1A}$ receptors in the hippocampus (Drevets et al., 2007), suggesting that a hyperactive HPA axis causes deficient serotonin activity. Another finding is that early experiences of stress in life sensitize the stress system, which contributes to a hyperactive HPA axis later in life (Heim et al., 2008). Research on neuroplasticity suggests a causal link between stress, impaired neuroplasticity, and depression (Pittenger & Duman, 2008). This research derives from findings that chronic stress impairs (e.g., synaptic and dendritic) neuroplasticity in brain areas implicated in depression (e.g., the hippocampus and PFC), depression is correlated with impaired neuroplasticity (e.g., reduced dendritic spine density of neurons in the hippocampus), and antidepressants improve neuroplasticity. Other mechanisms implicated in depression include circadian rhythms and the immune system (McClung, 2013).

Genetics research indicates that depression is moderately heritable, although evidence about genetic mechanisms is inconclusive. Twin studies produce heritability estimates around 37% (Sullivan, Neale, & Kendler, 2000). By contrast, bipolar disorder is one of the most heritable mental disorders, with heritability estimates around 90% (Kieseppä et al., 2004). Molecular genetic studies have targeted candidate genes related to serotonin. Some association studies indicate that risk for depression is influenced by genetic variation near the serotonin transporter gene, i.e., the serotonin-transporter-linked polymorphic region (5-HTTLPR) on chromosome 17. However, results have been inconclusive and GWAS have not independently implicated these genes (Sullivan et al., 2012). The largest GWAS on depression failed to identify any SNPs with genome-wide significance (Major Depressive Disorder Working Group of the Psychiatric GWAS Consortium, 2013). Authors of this study suggest that the null result reflects inadequate phenotypes due to adopting the DSM's broad definition of depression. Genetics and epidemiological research indicate that some subtypes of depression (e.g., recurrent depression, recurrent early-onset depression) are rarer and are more heritable (ibid, p. 502). This methodological point is significant because GWAS have only yielded significant associations for disorders (e.g., autism, bipolar disorder, schizophrenia) associated with low prevalence and high heritability (Wray et al., 2012). Hence, this null result likely reflects the heterogenous phenotypes produced by the

DSM's definition of depression, which lowered the statistical power (i.e., the probability of correctly rejecting a null hypothesis) of the GWAS to detect a significant association.

2.3 Some Mental Disorders are Constituted by Biological Mechanisms

The research reviewed above indicates that schizophrenia and depression are mental disorders underwritten by demonstrable biological mechanisms. Despite limited evidence regarding genetic mechanisms (cf. Schaffner, 2016, chs. 6–7), there is compelling evidence regarding the neurobiological mechanisms implicated in schizophrenia (e.g., dopamine, glutamate) and depression (e.g., serotonin, the HPA axis). Hence, Szasz's conclusion that abnormal psychological processes and behaviors cannot be explained in terms of demonstrable biological processes is false. This undercuts Szasz's argument—in its own (properly qualified) terms—that the explanatory concept of mental illness (2′) is invalid. It is certainly not an explanatory concept on par with 'deities' or 'witches.' This analysis also opposes the more general argument that the stereotypical signs of mental disorders are better explained by social mechanisms than biological ones.

My argument that mental disorders are underwritten by biological mechanisms emphasizes the importance of *pharmacological interventions* as a source of evidence (Tsou, 2012, 2017). This perspective draws on philosophical accounts that maintain that intervening with and manipulating theoretical (unobservable) entities in experimental contexts provides evidence for their reality and causal properties (e.g., see Hacking 1983; Franklin, 1996; Thagard, 1999, chs. 7–8; Woodward, 2003; Franklin & Perovic, 2019). The fact that the signs of mental disorders (e.g., psychosis) can be reliably treated or induced by pharmacological drugs—in predictable ways—provides *defeasible evidence* that these disorders are 'real entities' (or 'natural kinds') underwritten by neurobiological mechanisms. I articulate this view further in section 4. My argument assumes that a theory is well-confirmed or "robust" when multiple lines of partially independent research converge (or triangulate) on a common finding (e.g., see Wimsatt, 1981, 2007, ch. 7; Franklin & Howson, 1984; Culp, 1995; Stegenga, 2009; Soler et al. 2012; Schupbach, 2018). Accordingly, pharmacological interventions only constitute part of a larger assemblage of research (e.g., brain imaging studies, animal models) that contributes to the confirmation of a theory (cf. Radden, 2009, ch. 4). Hence, pharmacological interventions provide partial and fallible evidence that mental disorders arise from biological mechanisms.

While social explanations of mental disorders are not superior to biological explanations, this does not imply that social mechanisms (e.g., labeling, role adoption) play no causal role in mental disorders. In section 4.4, I argue that social mechanisms contribute to the specific *expression* of disorders in particular cultural contexts. On this view, biological mechanisms determine the general causal features of mental disorders (e.g., psychotic states, depressive states, manic episodes), while social mechanisms stabilize a more culturally-specific expression. Although I have argued that skeptics are incorrect to privilege social explanations over biological explanations, I have remained agnostic on the more specific issue of whether mental disorders are legitimately regarded as diseases. I address this issue in section 3.

3 Defining Mental Disorder

Definitions of mental disorder aim to delimit the proper objects of study in psychiatry. Boorse and Wakefield defend two of the most influential naturalistically-oriented accounts, which focus on articulating the *biological dysfunction* involved in mental disorders. Boorse maintains that mental disorders are conditions that interfere with the normal function of mental processes. Wakefield maintains that mental disorders are conditions caused by the failure of a mental mechanism to perform its naturally selected function, which harms individuals. A limitation of Wakefield's evolutionary account of dysfunction is its presupposition of uncritical 'adaptationist' assumptions. While Boorse offers a more promising account of biological dysfunction, Wakefield correctly stresses the need for a normative harm condition. Against Boorse and Wakefield, I argue that the naturalistic requirement of biological dysfunction should be weakened to the more general category of natural kinds. On this view, genuine mental disorders are biological kinds that lead to harmful consequences for individuals.

3.1 Boorse's Biostatistical Theory of Disease and Mental Disease

Christopher Boorse defends the most comprehensive and compelling naturalistic theory of disease, which rests on the assumption that "the normal is the natural" (Boorse, 1977, p. 554). The biostatistical theory (BST) maintains that health is (statistically) normal physiological functioning and disease (or pathology) is a state that interferes with normal functioning. Boorse (2014, p. 684) formulates the BST in four tenets:

1. A *reference class* is a natural class of organisms of uniform functional design, i.e., an age group of a sex of a species.
2. A *normal function* of an internal part or process within members of a reference class is its statistically typical contribution to survival or reproduction.

3. *Health* in a member of a reference class is *normal functional ability:* the readiness of each internal part to perform its normal functions on typical occasions with typical efficiency.
4. A *disease* or *pathological condition* is an internal state which impairs health, i.e., reduces one or more functional abilities below typical efficiency.

Boorse's theory is distinctive in its attempt to articulate concepts of health and disease that are *value-free* insofar as they are grounded in a purely descriptive (or factual) concept of biological function. This feature of the BST is motivated to identify an objective scientific basis for the theoretical concept of disease employed in medicine.

The BST is premised on a causal role account of function that maintains that the function of a trait is its statistically typical (normal) causal contribution to an organism's survival or reproduction.[13] Boorse's account emphasizes the goal-orientated (teleological) nature of biological organisms and assumes that species have a *normal functional design*:

> [T]he structure of organisms shows a means-end hierarchy with goal directedness at every level.... [T]he function of any part ... is its ultimate contribution to certain goals at the apex of the hierarchy... [T]he highest goals of organisms are indeterminate and must be determined by a biologist's interests... As a result ... different subfields of biology (e.g., genetics and ecology) may use different goals ... But it is only the subfield of physiology whose functions seem relevant to health. On the basis of what appears in physiology texts, I suggest that these functions are ... contributions to survival and reproduction.... Whatever goals are chosen, function statements will be value-free, since what makes a causal contribution to a biological goal is certainly an empirical matter. (Boorse, 1977, pp. 555–6)

This passage indicates the sense in which physiological functions can be specified in a value-free manner (cf. Engelhardt, 1976; Ereshefsky, 2009). Similarly, the *references classes* invoked in the BST are value-free because they are *factual descriptions* of naturally-occurring classes (or 'natural kinds') that exhibit distinctive functional designs within a species. Reference classes are not stipulated by conventional (or otherwise arbitrary) criteria, but *empirically discernible classes* distinguished by common functional design (cf. Cooper, 2005, ch. 1; Kingma, 2007). In medicine, Boorse (1977) maintains that the

---

[13] In the philosophical literature on biological functions, the two major contenders are: (1) *causal role accounts*: the function of a trait is its causal contribution to a larger system (e.g., see Cummins, 1975; Boorse, 1976b; Schaffner, 1993, ch. 8; Amundson & Lauder, 1994), and (2) *selected effect accounts*: the function of a trait is whatever it was naturally selected for (e.g., see Wimsatt, 1972; Millikan, 1989; Neander, 1991a, 1991b; Garson, 2019).

operative class is an age group of a sex. For Boorse, it is a putative *fact* that there are multiple reference classes (i.e., age classes within a sex) that correspond to distinctive concepts of health and disease (e.g., phenylketonuria, ovarian cancer, Alzheimer's disease).

Boorse's descriptive accounts of normal function and reference classes ground his value-free accounts of health and disease. Health is *normal functioning* within a reference class, wherein 'normal' is a statistical concept and 'functioning' is a biological concept. Boorse (2014) emphasizes that this concept of health is dynamic insofar as "normal function varies with both an organism's activity and its environment" (p. 685). Hence, health is *normal functional ability* (cf. Kingma, 2010) insofar as parts of organisms are *disposed* to perform normal functions with typical efficiency under appropriate occasions. Given the functional design of a reference class, there will be a *range* of normal values for most functions (e.g., heart rate, blood pressure) that contribute to survival and reproduction. Diseases are internal states that diminish normal functional abilities of an organism below typical efficiency (cf. Hausman, 2012). Hence, diseases are *subnormality* of physiological function relative to a reference class. On the assumption that health can be described in an empirical manner (e.g., specifying the functional design of various age classes within a sex), pathological conditions are empirically discernable internal states (e.g., narrowed or blocked arteries, insufficient insulin production) that impair statistically normal functioning.

While Boorse agrees with Szasz that genuine diseases should be identifiable in an objective (value-free) manner, he rejects Szasz's conclusion that mental disorders are not diseases. Boorse (1976a) argues that, so long as there are normal functions of mental (or psychological) processes that contribute to survival or reproduction, there is no reason to deny the existence of mental diseases:

> If certain types of mental processes perform standard functions in human behavior, it is hard to see any obstacle to calling unnatural obstructions of these functions mental diseases. . . . Perceptual processing, intelligence, and memory clearly serve to provide information about the world that can guide effective action. Drives serve to motivate it. Anxiety and pain function as signals of danger. (p. 64)

With respect to what differentiates physical from mental health, Boorse maintains that physical health involves physiological functions, while mental health is the special case that focuses on the normal functions of mental processes. Boorse (1976a) argues that this view only assumes that: (1) some mental processes (e.g., perception, memory) play a causal role in action, and

(2) these mental processes play standard functional roles in a species.[14] In this framework, Boorse (1976a) argues that some mental disorders are pathological conditions that interfere with normal mental functioning:

> It seems certain that a few of the recognized mental disorders are genuine diseases, whether mental or physical. Even without knowledge of the relevant functional systems, one can sometimes infer internal malfunction immediately from biologically incompetent behavior. . . . Some mental patients, e.g., catatonic schizophrenics, are clearly incompetent with respect to . . . biological goals . . . One should not underestimate the mileage that can be got out of elementary functional assumptions . . . We may surely assume . . . that the main function of perceptual and intellectual processes is to give us knowledge of the world. . . . [I]f my cognitive functions are disrupted to a highly unusual degree . . . it seems safe to call my condition an unnatural dysfunction, i.e., a disease. By this standard, schizophrenia and all other psychoses with thought disorders look objectively unhealthy. Moreover, if one accepts the traditional [psychoanalytic] descriptions of the neurotic process, very limited functional assumptions will suffice to construe serious neurosis as a disease. (pp. 76–7)

If the normal function of perceptual and cognitive systems is to provide information about the environment, then severe impairments to this function (e.g., psychosis) render schizophrenia a mental disease. Boorse's analysis of neurosis is less compelling given his assumption of certain psychoanalytic assumptions (e.g., that a 'conflict-resolution mechanism' is functioning incorrectly). However, there are plausible hypotheses regarding the interferences of normal functioning involved in these conditions, e.g., anxiety involves a disruption of the fear response. To have a unified theory of mental health, Boorse recommends that psychiatry should articulate a detailed theory of the functions of the normal human mind, which would provide an objective foundation for identifying mental diseases (cf. Washington, 2016).

3.2 Wakefield's Harmful Dysfunction Analysis of Mental Disorder

In contrast with Boorse's purely naturalistic account of disease, Jerome Wakefield (1992a, 1992b) defends the most influential hybrid account of mental disorder, which invokes both naturalistic and normative factors. According to Wakefield's harmful dysfunction (HD) analysis, a mental disorder must satisfy two criteria:

---

[14] Boorse's (1976a) view of 'mental functions' does not imply metaphysical dualism, and he argues that a functionalist version of identity theory (Putnam, 1960; Davidson, 1970) can satisfy these two conditions.

(1) Dysfunction: the condition results from the failure of a mental mechanism to perform its naturally selected function.
(2) Harm: the condition causes significant harm to the person as judged by current cultural standards.

The HD definition implies that mental disorders involve a factual aspect (dysfunction) and an evaluative aspect (harm). Like Boorse, Wakefield regards judgments of dysfunction to be value-free and determined in a purely scientific matter. Whereas Boorse's BST presents biological dysfunction as a necessary and sufficient condition for disease, Wakefield's HD analysis presents harm and dysfunction as necessary and jointly sufficient conditions. Wakefield (1992b) argues that the HD account offers the "correct analysis" insofar as it avoids the conceptual problems facing alternative accounts of disease and disorder (p. 373).

The dysfunction invoked in Wakefield's HD analysis is an evolutionary concept. In contrast with the causal role account of function adopted by Boorse, Wakefield adopts a selected effect account of function (see note 13), wherein dysfunction is the failure of an internal mechanism to perform its naturally selected function. For Wakefield (1992b), the natural function of a mechanism (e.g., the heart, perceptual mechanisms) is a *non-accidental effect* of the mechanism (e.g., pumping blood, providing information about the environment) that can explain the existence, structure, or activity of the mechanism. Wakefield (1992b) assumes that natural selection provides the best explanation for why organisms' internal mechanisms are designed in such beneficial ways:

> [T]hose mechanisms that happened to have effects on past organisms that contributed to the organisms' reproductive successes over enough gener-ations increased in frequency and hence were "naturally selected" .... Thus, an explanation of a mechanism in terms of its natural function may be considered a roundabout way of referring to a causal explanation in terms of natural selection. Because natural selection is the only known means by which an effect can explain a naturally occurring mechanism that provides it, evolutionary explanations presumably underlie all correct ascriptions of natural functions. (p. 383)

This account of natural function, according to Wakefield (1992b), is applicable to the natural functions of *mental mechanisms*:

> [M]ental mechanisms, such as cognitive, linguistic, perceptual, affective, and motivational mechanisms, have such strikingly beneficial effects and depend on such complex and harmonious interactions that the effects cannot be entirely accidental. Thus, functional explanations of mental mechanisms

are sometimes justified by what we know about how people manage to survive and reproduce. For example, one function of linguistic mechanisms is to provide a capacity for communication, one function of the fear response is to help a person to avoid danger, and one function of tiredness is to bring about rest and sleep. These functional explanations yield ascriptions of dysfunctions when respective mechanisms fail to perform their functions, as in aphasia, phobia, and insomnia, respectively. (p. 383)

Wakefield's account of the function of mental mechanisms implies that dysfunction is the failure of a mental mechanism to perform its naturally selected function. Although Wakefield acknowledges that identifying the natural function of mental mechanisms is difficult and subject to debate, discovering the function and dysfunction of mental mechanisms is a factual matter.

The normative harm requirement in Wakefield's HD analysis implies that dysfunctional conditions that lead to harm are mental disorders. Wakefield maintains that harm is a commonsense and practical ('folk') concept. He includes this requirement as a necessary condition for disorder because there is often a disconnection between dysfunction and harm. One reason for this disconnection is a difference between the environment of selection and current environments. For instance, high levels of male aggression were beneficial in primitive environments, but impairment of this function is not currently harmful. Another reason is that small decreases in reproductive fitness may be important over long evolutionary time-scales, but not necessarily harmful in the practical sense relevant for psychiatry. As Wakefield (1992b) puts it: "The mental health theoretician is interested in the functions that people care about and need within the current social environment, not those that are interesting merely on evolutionary theoretical grounds" (p. 384).

Wakefield (1992b) contrasts his account to "biological disadvantage" accounts (Kendell, 1975; Boorse, 1976a, Scadding, 1990) that conceptualize mental disorder as dysfunctional conditions that compromise survival or reproduction. These authors proffer value-free accounts of mental disorder by implicitly equating compromised survival or reproduction with harm (cf. Boorse, 1977, p. 544). Wakefield (1992b) complains that this equation fails to yield the concept of harm that is relevant in medicine and psychiatry:

A condition can reduce fertility without causing real harm; marginally lowered fertility is serious over the evolutionary time-scale, but it may not affect an individual's well-being if the capacity for bearing some children remains intact. And some serious harms, such as chronic pain or loss of pleasure, might not reduce fertility or longevity at all… [T]here are many harmful conditions, such as postherpetic neuralgia and psoriasis, that are clear cases of disorder but have no effect on mortality or fertility. This is likely to be even more true of mental disorders. It would seem that the harm requirement must be added to, rather than derived from, the evolutionary requirement. (p. 378)

Given the independence of detriments to *biological fitness* (i.e., the ability to survive and reproduce) and what is deemed harmful by current cultural standards, Wakefield argues that the harm requirement must be included as a separate necessary condition. Hence, a definition of disease or disorder that will be of practical relevance to doctors and psychiatrists needs to incorporate the (sometimes contentious) evaluative judgments made in medicine and psychiatry regarding when an individual's capacity to live a normal life is significantly harmed.

3.3 Natural Function, Biological Dysfunction, and Adaptationism

Boorse and Wakefield both identify the naturalistic basis of mental disorders with biological dysfunction; however, their accounts of biological dysfunction rest on different views on biological function. Boorse's BST assumes a causal role concept of function, wherein a natural function amounts to the (statistically) normal contribution that an internal part contributes to an organism's survival or reproduction. Wakefield's HD analysis assumes a narrower, selected effect concept of function, wherein a natural function is the beneficial ('adaptive') effect of a mental mechanism that was naturally selected. Boorse's account implies that a functional part must contribute to current fitness, whereas Wakefield's account implies that a functional mechanism must have contributed to fitness in the past and been naturally selected for that effect (Boorse, 2014, p. 687).

Wakefield's account of dysfunction, and his selected effect account of function in particular, presupposes oversimplified ('Panglossian') *adaptationist* assumptions criticized by philosophers of evolutionary biology (Gould & Lewontin, 1979; Lewontin, 1979; Lloyd, 1999). Gould and Lewontin (1979) present the "adaptationist programme" as a deeply ingrained style of reasoning prevalent in sociobiology and evolutionary psychology, wherein organisms are organized into traits and adaptationist stories are told to explain the existence of these traits as being optimally designed by natural selection. This evolutionary style of reasoning is problematic in its: (1) failure to distinguish current utility of a trait with reasons for its origin, (2) uncritical application of adaptationist stories, (3) reliance on plausibility alone as a criterion for theory acceptance, and (4) failure to consider alternative reasons for the existence of traits. Gould and Lewontin complain that adaptationist explanations are often untestable in the face of evidence since new ('just-so') adaptationist stories can always be told. This captures the sense in which consistency of an adaptationist story with natural selection ('plausibility alone') is the sole criterion for theory acceptance. In place of the adaptationist programme, Gould and Lewontin (1979) recommend a pluralistic view of evolutionary change that recognizes alternative explanations for the emergence of traits (pp. 590–3):

(1) Non-adaptive traits are entrenched due to contingent factors (e.g., genetic drift, demographic events).
(2) A non-adaptive trait is a correlated consequence of selection directed elsewhere (e.g., pleiotropy).
(3) A trait is favored by selection, but is non-adaptive because of the current environment, or an organism is well adapted to the environment, but this 'good design' is due to phenotypic plasticity, not natural selection.
(4) Organisms develop multiple adaptive trait solutions to problems, but no solution is more optimal than another.
(5) A beneficial trait (i.e., 'spandrel') is an accidental bi-product of other traits, rather than a direct product of natural selection.

While Gould and Lewontin (1979) acknowledge that natural selection is the most important mechanism for explaining evolutionary change, they argue that exclusive focus on adaptationist explanations and neglect of alternative explanations has resulted in an impoverished paradigm.

Wakefield's (1992b) account of the natural function of mental mechanisms is adaptationist in its assumption that "natural selection is the only known means by which an effect can explain a naturally occurring mechanism that provides it" (p. 383). Lilienfeld and Marino (1995) argue that many mental functions are not adaptations in this sense and many higher-order mental traits (e.g., language, mathematical ability) are likely to be *exaptations*, rather than adaptations. Exaptations are traits that now enhance fitness, but were not directly designed by natural selection for their current function (Gould & Vrba, 1982). Gould and Vrba (1982) distinguish between two types of exaptations (p. 5):

(1) *Secondary adaptations*: traits that were naturally selected for a particular function (i.e., adaptations), but were subsequently co-opted for another use. For example, bird feathers were originally selected for thermoregulation, but subsequently took on the function of flight.
(2) *Spandrels*: traits whose origin cannot be ascribed to the direct action of natural selection, but currently serve some beneficial function. For example, delayed ossification of skull bones in human infants has the benefit of allowing relatively large fetal heads to travel through narrow

birth canals; evidence that this trait is not an adaptation is that mammalian relatives, who are born in eggs, share this trait (Gould, 1991).

The existence of spandrels goes much further in questioning adaptationist orthodoxy because many currently beneficial traits might have arisen for entirely non-adaptive reasons (Gould, 1991). If many higher-order mental functions—especially those that are impaired in the domain of abnormal psychology—are spandrels rather than adaptations, then Wakefield's adaptationist account of mental function provides a spurious ground for an account of dysfunction.[15]

Boorse's causal role account of function is more philosophically promising as a foundation for an account of biological dysfunction than Wakefield's selected effect account (cf. Garson, 2019, ch. 8). First, it offers a broader view of biological dysfunction that can include conditions other than those that involve failures of mechanisms to perform their naturally selected function. While Wakefield's account of natural function undoubtedly individuates a class of dysfunctional conditions, it is unclear why we should restrict mental disorders to conditions that involve evolutionary dysfunctions of this sort. Standard medicine certainly does not make such a restriction (Murphy, 2020a). Boorse's causal role account is more promising because the natural function of a trait is viewed more broadly as its typical causal contribution to an organism's biological fitness, which more accurately captures the concepts of function and dysfunction implicit in medicine. Second, Boorse's causal role account is not committed to speculative adaptationist assumptions regarding the origins of mental mechanisms. In this regard, Boorse (2014) notes that an advantage of his account, compared to Wakefield's, is that "no evolutionary evidence is necessary to establish a trait's function" (p. 687). For instance, if mechanisms underwriting mood regulation turn out to be spandrels (rather than adaptations), Boorse could maintain that some impairments in mood regulation are dysfunctions, whereas Wakefield could not. Finally, in its appeal to statistical normality, the BST offers an account of function and dysfunction that lends itself more readily to empirical analysis, as opposed to speculative evolutionary theorizing.

---

[15] Against the argument that the HD account rules out disorders (e.g., acalculia, dyslexia) that involve the disruption of higher-order mental traits (e.g., calculation, reading) that were not naturally selected (Lilienfeld & Marino, 1995), Wakefield (1999) argues that his account only requires dysfunction of *any* naturally selected mental mechanism, which can *indirectly* disturb these higher-order traits. Lilienfeld and Marino (1999) point out that if harms do not *directly* result from dysfunction, the HD account is overinclusive since it would imply that individuals would have 'dressing disorder' or 'driving disorder' if they went blind or lost use of their limbs. For further discussion of evolutionary explanations of mental disorders, see Buss et al. (1998); Murphy and Stich (2000); Murphy and Woolfolk (2000); Wakefield (2000); Murphy (2006, ch. 8); Varga (2012); and Garson (2019, ch. 11).

3.4 Harm, Values, and Philosophical Debates about Definitions

Another apparent point of contention between Wakefield and Boorse is the issue of harm. Whereas Boorse argues that mental disease can be defined in a purely factual manner, Wakefield argues that a definition of mental disorder needs to incorporate evaluative judgments of harm. On the issue of values, Boorse acknowledges the need for value-judgments in *practical contexts* of medicine and psychiatry, and he draws a distinction between theoretical (value-free) and practical (value-laden) concepts of disease. In earlier formulations, Boorse (1975, 1976a) distinguished between "disease" and "illness," arguing that the concept of disease is factual, while the concept of illness requires value-judgments. In particular, illnesses are *harmful diseases* that satisfy further normative criteria (Boorse, 1976a, p. 63):

> A disease is an *illness* only if it is serious enough to be incapacitating, and therefore is
>
> (i)   undesirable for its bearer;
> (ii) a title for special treatment; and
> (iii) a valid excuse for normally blameworthy behavior

Boorse's definition of "illness" (and "mental illness") acknowledges the necessity of evaluative judgments and takes the same general form as Wakefield's hybrid definition of "disorder" (and "mental disorder"). Boorse (1997) subsequently reformulated this distinction as a demarcation between the theoretical concept of "disease" and various practical "disease-plus" concepts (e.g., "treatable disease," "disabling disease"). Hence, Boorse recognizes the need for value-laden concepts of disease in practical contexts. His insistence that disease is a value-free concept ultimately stems from his goal to define the "the technical usage of 'disease' found in textbooks of medical theory" (Boorse, 1975, p. 60).[16] By contrast, Wakefield's goal is to define a concept of disorder that is "fundamental in theory and practice in the mental health field" (Wakefield, 1992b, p. 373), which leads to his conclusion that disorder is "a practical concept that is supposed to pick out only conditions that are undesirable and grounds for social concern, and ... no purely scientific nonevaluative account ... captures such notions" (Wakefield, 1992a, p. 237). Given Boorse's acknowledgment that disease-plus concepts require evaluative judgments, the issue of harm is a red-herring. The

---

[16] Boorse aims to defend the scientific objectivity of medicine by demonstrating that it rests on value-free concepts (cf. Kious, 2018). From the perspective of philosophy of science, this motivation is somewhat peculiar. Most contemporary philosophers of science assume that values are ubiquitous in science, but the presence of values does not necessarily compromise scientific objectivity (Longino, 1990; Kincaid, Dupré, & Wylie, 2007; Douglas, 2009; Padovani, Richardson, & Tsou, 2015; Reiss & Sprenger, 2020).

conflict between Boorse and Wakefield stems from the fact that they are defining incommensurable *explananda*.

Peter Schwartz (2007a, 2014) provides a compelling argument against the pervasive use of conceptual analysis to assess the *correctness* of definitions of disease. Conceptual analysis is an *a priori* method for assessing the adequacy (or correctness) of definitions of concepts and is strongly entrenched in the history in philosophy. In the 'classical theory of concepts,' a concept is defined by necessary and sufficient criteria, and the correctness of definitions is evaluated (via philosophical intuition and thought experiments) in terms of how well they accommodate putative cases and avoid (actual or possible) counterexamples.[17] Conceptual analysis is related to the different *types* of definitions that philosophers defend (Gupta, 2019):

(1) *stipulative definitions*: stipulate meanings to terms, without considering prior or established meanings of terms.
(2) *descriptive definitions*: specify meanings to terms that are adequate to established usages.
(3) *explicative definitions*: specify meanings to terms that respect some central uses of a term, but is stipulative on others.

Conceptual analysis is assumed to be appropriate for assessing the *correctness* of (2). By contrast, (1) and (3) are assumed to be evaluated as more or less *useful* (rather than correct) on the basis of pragmatic criteria (e.g., explanatory power, simplicity). Schwartz argues that debates about disease are marred by the excessive use of conceptual analysis, which is assumed to be an effective means for evaluating the correctness or discovering the 'true meaning' of definitions (cf. Lemoine, 2013). Against this assumption, Schwartz (2007a) observes that, although these issues have been debated for over 60 years (e.g., debates between naturalists and normativists), virtually no progress or consensus has been achieved. In this regard, Schwartz (2007a) argues that Boorse's and Wakefield's analyses, which offer descriptive definitions of disease and disorder, are problematic insofar as they assume that conceptual analysis is a reliable means for evaluating the *correctness* of their definitions. More generally, Schwartz (2014) maintains that attempts to define 'disease' or other concepts "should avoid any claim to be discovering the meaning or correct criteria of application of these concepts" (p. 573).

---

[17] Conceptual analysis has been criticized by naturalistically-oriented philosophers, who object to its *a priori* nature, reliance on intuitions, appeal to folk concepts, and misleading view of meaning (e.g., see Ramsey, 1992; Stich, 1992; DePaul & Ramsey, 1998; Laurence & Margolis, 2003). For defenses of conceptual analysis, see Chalmers (1996) and Jackson (1998); Machery (2017) defends a naturalistic approach to conceptual analysis.

As an alternative to conceptual analysis, Schwartz (2007a, 2014) recommends the more deflationary approach of *philosophical explication*—defended by Carnap (1950, ch. 1) and Quine (1960, ch. 7)—as a more promising philosophical perspective for understanding and guiding debates about definitions of concepts. While Carnap's and Quine's views differ (e.g., Carnap views explication as a clarification of pre-theoretical concepts, while Quine views it as a form of elimination), they both take explicative definitions to be *new (stipulative) definitions of old (descriptive) definitions*. Schwartz (2014) finds this viewpoint attractive insofar as definitions are "evaluated as proposals about how to define a term in the future, not as discoveries about the current meaning or criteria of application" (p. 573). Since explicative definitions may change the meaning of established terms, conceptual analysis is less relevant for evaluating the adequacy of definitions. Moreover, since proposed explicative definitions are evaluated on pragmatic grounds given the specific purposes for which definitions are formulated, "no single concept can be relied upon in all situations; there may need to be different definitions for different contexts" (Schwartz, 2007a, p. 60). From this perspective, longstanding debates regarding the meaning or nature of disease can be recast as arguments regarding the advantages and disadvantages of different accounts: "Choosing an account becomes not so much a hunch about which theory is correct but instead a choice of which theory to clarify and apply" (Schwartz, 2007a, p. 60). As suggested above, the apparent conflict between Boorse's and Wakefield's accounts on the issue of harm is not a substantive disagreement, but reflects the different interests motivating their accounts.

3.5 Proposal: Mental Disorders are Biological Kinds with Harmful Effects

In the spirit of philosophical explication, I propose that mental disorders are natural (i.e., biological) kinds that lead to harmful consequences. Since my interest in defining mental disorder is to clarify the proper objects of study *and treatment* in psychiatry, the intended explanandum of this definition is closer to Wakefield's insofar as it is a concept that is relevant to both researchers and practicing clinicians. Accordingly, I am sympathetic to Wakefield's argument that a definition of mental disorder requires a normative component (cf. Powell & Scarffe, 2019). Wakefield (1992b) argues that a disorder is harmful when it "causes some harm or deprivation of benefit to the person as judged by the standards of a person's culture" (p. 384). This concept can be reformulated more precisely as follows: a disorder is harmful when it *compromises the capacity of a person to live a normal or unimpeded life* as judged by cultural standards. Moreover, harms must be *direct consequences* of features (e.g. psychotic episodes) of biological kinds, rather than indirect consequences of socially harmful environments (e.g., discrimination against individuals exhibiting these traits). This normative criterion is typically appealed to when judging whether a condition merits treatment or intervention. Hence, mental disorders are biological kinds that compromise the capacity of an individual to lead an unimpeded life.

While I argued that Boorse's account of biological dysfunction is more promising than Wakefield's, I contend that being a biological kind is a more pragmatically useful candidate for the naturalistic (or factual) component of mental disorders. This aspect of my proposal is more stipulative than Boorse's and Wakefield's accounts insofar as it deviates from the common understanding of mental disorders as *pathological* conditions. One motivation for this shift in understanding is to defend a naturalistic standard that is readily identifiable for individual disorders. A weakness of the accounts of biological dysfunction recommended by the BST and HD analysis is that they present naturalistic standards that are not easily ascertainable in practice. For the HD account, this difficulty is related to obtaining evidence for the evolutionary origins of the psychological functions that are impaired in disorders. For the BST, the difficulty concerns identifying the proper threshold for *subnormality*; Schwartz (2007b) aptly characterizes this difficulty as the 'line-drawing problem.' Deflating the requirement of biological dysfunction to biological kinds shifts the attempt to identify biological dysfunction to the search for biological mechanisms. My discussion of the biological mechanisms underwriting schizophrenia and depression in section 2.2 demonstrates that this is an attainable standard.

Another pragmatic motivation for this stipulation is that it can accommodate the variety of conditions that are treated by mental health professionals, which is a goal typically emphasized in normative accounts. My proposed account aims to be descriptively adequate in this sense. The requirement that mental disorders are biological kinds is broad enough to include disorders (e.g., schizophrenia, bipolar disorder) that are caused by dysfunctional biological mechanisms. However, certain conditions (e.g., acute depression or anxiety) might turn out to be underwritten by biological mechanisms that behave in predictable ways, but fall within the (statistically) normal range of biological functioning (cf. Maung, 2016; Stegenga, 2018, ch. 4). My account implies that they are genuine mental disorders that merit treatment, so long as they result in harmful consequences. The fact that Boorse's and Wakefield's accounts would rule out these conditions, speaks against their desirability. The naturalistic standard of being a biological kind is sufficiently liberal to accommodate the broad range of conditions treated by mental health professionals. Conversely, it is sufficiently restrictive to ensure that genuine mental disorders have a *common biological basis*, which renders classifications of such kinds predictively useful in treatment contexts. I articulate this argument more comprehensively in section 4.

4 Natural Kinds in Psychiatry

Some mental disorders (e.g., schizophrenia, depression) are natural kinds (i.e., homeostatic property cluster kinds): classes of abnormal behavior whose characteristic signs are constituted by sets of stable biological mechanisms. This view can explain the relative stability of psychiatric kinds and the projectable inferences yielded by their classifications. Against Hacking's argument that the objects of classification in psychiatry are 'moving targets' because of the 'looping effects' of human kinds, biological kinds remain stable in spite of classificatory feedback. Hacking's analysis highlights the ways in which social mechanisms (e.g., imitation of stereotypes, role adoption) can causally influence the expression of mental disorders. I examine—with reference to cross-cultural research, transitory mental disorders, and culture-bound syndromes—ways in which biological and social mechanisms interact to influence the stability and expression of mental disorders. I conclude that the proper objects of study and classification in psychiatry are biological kinds.

4.1 Natural Kinds in the Human Sciences, Projectability, and Essentialism

The distinction between natural and artificial kinds is intended to demarcate classes that are discovered versus invented, respectively (Bird & Tobin, 2018). Natural kinds are *naturally occurring classes* (e.g., electrons, quarks, H_2O) that are discovered by classifiers and reflect the causal structure of the world. By contrast, artificial kinds are *conventional classes* (e.g., furniture, skyscrapers, books) that are invented and reflect the interests of classifiers. Hence, natural kinds are natural (or 'real') classes that constitute the proper objects of scientific study (e.g., fundamental physical particles, chemical elements). Artificial kinds are socially constructed classes that do not reflect naturally-occurring divisions in nature. On the assumption that *all* scientific classifications involve elements of human convention, the distinction between natural kinds and artificial kinds is best understood as a distinction in degree, rather than a sharp distinction in kind. While there is no privileged or uniquely correct way of classifying kinds (Dupré, 1993, 2000), some classifications are *more natural* than others insofar as they more accurately individuate classes distinguished by natural properties. With respect to human classifications, some classifications (e.g., 'female,' 'introvert,' 'hemophilia') are more natural than others (e.g., 'liberal,' 'teacher,' 'widow') insofar as they individuate classes of people in terms of natural (i.e., biological) properties common to their members.

Philosophical accounts of natural kinds, following a tradition initiated by Goodman (1955) and Quine (1969, ch. 5), are motivated to explain how natural kinds terms (or predicates) yield projectable inferences. In this tradition, a satisfactory account of natural kinds should explain how natural kind terms yield reliable and non-trivial (i.e., ampliative) predictions about kind members on the basis of induction. Essentialist accounts maintain that natural kind terms (e.g., 'water') are projectable because members of a kind share intrinsic natural properties that participate in laws in nature (Khalidi, 2013, ch. 1). In particular, there are a set of necessary and sufficient natural properties that constitute the essence of a kind (e.g., the molecular structure of H_2O). Reliable predictions yielded by kind terms (e.g., water will freeze at $0°C$) are explained by the intrinsic natural properties shared by kind members. While essentialist accounts provide a straightforward explanation of the projectability of natural kind terms, they are not well-suited for describing the messier classes studied in the special sciences (e.g., biology, medicine, psychology) insofar as these classes (e.g., species, diseases) are neither immutable nor constituted by exceptionless laws of nature (cf. Cartwright, 1983).

Richard Boyd (1991, 1999a) defends a non-essentialist theory of natural kinds—i.e., homeostatic property cluster (HPC) kinds—that is formulated primarily to explain how classifications in special sciences (e.g., biology) yield projectable inferences. The key features of HPC kinds are the following (Boyd, 1999a, pp. 143–4):

(1) There is a family of properties (F) that are contingently clustered in nature.
(2) Their co-occurrence is the result of "homeostasis": either the presence of some properties tends to favor the presence of others, or there are underlying mechanisms that tend to maintain the properties in F, or both.
(3) There is a kind term (t) that is applied to things in which the homeostatic clustering of most of the properties in F occurs.

In Boyd's theory, the capacity of a kind term to adequately represent relevant causal structures—i.e., the properties and mechanisms referred to in (2) that *cause properties to cluster in a regular and non-accidental way*—is what explains successful projectable inferences. More precisely, Boyd maintains that successful projectable inferences are explained by an accommodation (or fit) between our classifications and relevant causal structures. According to the 'accommodation thesis,' successful inductive inferences and explanatory generalizations that are generated within a paradigm offer (abductive) evidence that the posited kinds are representing (or accommodating) genuine causal regularities in the world. As Boyd (1991) puts it: "Kinds useful for induction or explanation must always 'cut the world at its joints' in this sense" (p. 139).

Boyd's account of HPC kinds has been influential in philosophy of psychiatry, and various accounts of mental disorders as HPC kinds have been defended (e.g., Beebee & Sabbarton-Leary, 2010; Kendler, Zachar, & Craver, 2011; Tsou, 2013, 2016). In contrast to Boyd's neutrality regarding the properties and mechanisms underwriting HPC kinds (Boyd, 1991, 1999a), I defend a qualified essentialist interpretation of Boyd's theory that requires that HPC kinds are underwritten by some *intrinsic* (i.e., biological) properties or mechanisms.[18] In the literature on HPC kinds, there is disagreement regarding whether *any* of the properties or mechanisms underwriting HPC kinds need to be intrinsic. Philosophers of biology who discuss species as HPC kinds emphasize the importance of relational properties and mechanisms (e.g., phylogenetic relations, interbreeding with conspecifics, exposure to similar environmental pressures) that maintain the stability of property clusters associated with a species (Boyd 1999a; Wilson, Barker, & Brigandt 2007; Ereshefsky 2017). Some (e.g., Griffiths 1999; Millikan, 1999) argue that the cluster of properties associated with species can be explained *exclusively* in terms of relational properties (e.g., descent from a common ancestor), while others (e.g., Boyd 1999b, Wilson 1999) insist that *some* intrinsic properties (e.g., genetic properties) are necessary.[19]

To yield robust projectable inferences, HPC kind terms need to individuate *some* intrinsic properties or mechanisms. This position follows the essentialist assumption that the intrinsic properties shared by kind members explains the projectability of their classifications. In this regard, it is important to distinguish two distinctive roles that causal mechanisms play in the theory of HPC kinds, which are sometimes conflated in the literature:

---

[18] Whereas Boyd (1999a) adopts the *a posteriori* methodological stance that the 'naturalness' of natural kinds is whatever reference to such kinds (including social kinds) contribute to the accommodation of classificatory schemes to causal structures (pp. 158–9), my interpretation of HPC kinds identifies the 'naturalness' of natural kinds with *intrinsic natural* (i.e., biological) *properties* (cf. Craver, 2009). This standard renders the theory of MPC kinds (Kendler et al., 2011) essentialist insofar as it appeals to intrinsic (e.g., neurobiological) mechanisms. Beebee and Nigel-Sabbarton (2010) avoid the issue of essentialism and usefully distinguish various philosophical senses of the term.

[19] In responding to an argument by Millikan (1999) that species are 'historical kinds' maintained by relational mechanisms (e.g., a copying process), Boyd (1999b, p. 81) emphasizes that species are HPC kinds underwritten by *both* intrinsic mechanisms (e.g., genetic properties) and relational mechanisms.

(1) From a metaphysical standpoint, they fix the stability of natural classes and explain the unity ('homeostasis') of the properties that are used to identify kinds.

(2) From an epistemological standpoint, they ground the stability of projectable inferences made about such kinds.

While relational properties and mechanisms can address (1), they cannot address (2). This is because relational properties (e.g., descent from a common ancestor, interbreeding with conspecifics) are *too general* to yield specific predictions about members of a particular species (cf. Devitt, 2008). To provide projectable inferences about particular species members, a kind term needs to individuate *some* intrinsic properties shared by its members (e.g., genetic properties that explain the physiological and morphological features of species). With respect to mental disorders, this implies that a kind term (e.g., 'schizophrenia') needs to individuate some intrinsic (e.g., neurobiological) properties shared by its members to yield projectable inferences. Hence, HPC kinds in psychiatry are biological kinds constituted by 'partly intrinsic biological essences' (Devitt 2008, 2010), although these essences are constituted by sets of interacting biological mechanisms and permit exceptions and variability.

4.2 Hacking on Looping Effects and the Instability of Human Kinds

Ian Hacking (1995b, 1999) argues that the kinds studied in the human and social sciences ('human kinds,' 'kinds of people,' or 'interactive kinds') are fundamentally different from the natural kinds studied in natural sciences such as physics and chemistry. In particular, Hacking (1995b) argues that the 'looping effects' of human science classifications render the kinds (i.e., objects of classification) studied in the human sciences inherently unstable. Therefore, human science classifications are incapable of yielding stable projectable inferences.

A distinctive feature of objects of classification in the human sciences is that the people being classified are *aware of and will change in response to how they are classified*. While objects of classification in the natural sciences (e.g., electrons, water) are *indifferent* to how they are classified, objects of classification in the human sciences (e.g., children with autism, individuals with schizophrenia) *interact* with their classifications (Hacking, 1999, ch. 4). This distinctive feature of classified people gives rise to the looping effects of human kinds. Looping effects are a phenomenon wherein the meaning of a human science classification (e.g., 'multiple personality') changes the experiences and behaviors of classified people (e.g., individuals act in accordance with prevailing stereotypes) such that the classification must be revised to accommodate such changes. As Hacking (1995a) puts it: "*People classified in a certain way tend to conform to or grow into the ways that they are described*; but they also evolve in their own ways, so that the classifications and descriptions have to be constantly revised" (p. 21).

Hacking's argument that human science classifications change the people they classify motivates his further contention that the objects of classification (i.e., 'natural kinds' or 'indifferent kinds') in the natural sciences are stable, whereas the objects of classification (i.e., 'human kinds' or 'interactive kinds') in the human sciences are inherently unstable (Hacking 1999, p. 108). Hacking (2007) writes:

> We think of ... kinds of people ... as definite classes defined by definite properties. ... But it is not quite like that. They are moving targets because our investigations interact with the targets themselves, and change them. And since they are changed, they are not quite the same kind of people as before. The target has moved. That is the looping effect. (p. 293)

Hacking draws the radical conclusion that—because human science classifications change the people they classify—objects of human science classifications ('kinds of people') are perpetually unstable entities. Since the meaning of human science classifications alter the people they classify, classified people will constantly change in response to how they are classified and classifications will need to be constantly revised to accommodate such changes. This is the precise sense in which Hacking suggests that the looping effects of human science classifications render their objects of classification 'moving targets.' If Hacking's analysis is correct, human science classifications cannot yield any robust projectable inferences given the inherent instability of the objects they aim to classify (Hacking, 1995b).

Tsou (2007, 2013, 2016) argues that Hacking commits a hasty generalization in concluding that the objects of classification in the human sciences are inherently unstable. In particular, Hacking's conclusion is a non-sequitur that does not follow from his (correct) claim that human science classifications change the people they classify. Hacking's argument can be reconstructed as follows:

(1) People are aware of and will change in response to how they are classified (*the reflexivity of human subjects*).

(2) Human science classifications change the experiences and behavior of the people they classify (*classificatory feedback*).

(3) Human science classifications must be revised in order to accommodate changes that *the classifications cause among classified people* (*looping effects*).

(4) Thus, objects of classification in the human sciences are inherently unstable (*human kinds are moving targets*).

While (1) and (2) are undoubtedly true, Hacking fails to demonstrate that these premises entail (3), which is the key premise that establishes (4). Tsou (2007) argues that Hacking's analysis conflates (2) and (3).[20] Classificatory feedback is a ubiquitous feature of most human science classifications. So long as people being classified are aware of and will change in response to how they are classified, classificatory feedback will be present.[21] However, the mere presence of classificatory feedback will not necessarily imply looping effects. While diagnosing an individual as 'depressed' will invariably change the experiences and behavior of the individual, the ways in which classificatory feedback change groups of classified people will not necessarily require changing the meaning of 'clinical depression,' or require revisions in the criteria used to identify depression (e.g., persistent feelings of sadness). For looping effects to occur, classificatory feedback must change the experiences and behavior of individuals—*in a uniform manner for a significant number of individuals who fall under that classification*—such that the *definitive characteristics* of that classification (e.g., the criteria that constitute membership for a classification) must be revised. While there are examples of looping effects in the human sciences (e.g., multiple personality), their prevalence is far less frequent than Hacking suggests.

Against Hacking's contention that the objects of human science classifications are inherently unstable ('moving targets'), HPC kinds are stable objects of classification *despite the presence of classificatory feedback*. If a human science classification refers to an HPC kind with a partly intrinsic (biological) essence, then classificatory feedback will not render that object of classification (i.e., a biological kind) unstable. This argument clearly applies to medical classifications (e.g., 'HIV,' 'chlamydia,' 'breast cancer') that can stigmatize patients. While being diagnosed as HIV positive will inevitably change the experiences and behavior of diagnosed individuals, the ways in which individuals change will not require revisions to the meaning of HIV or the symptoms used to diagnose it. The stability of this object of classification is explained entirely by the fact that HIV is an HPC kind constituted by a partly intrinsic biological essence (i.e., it is an immunodeficiency virus). Classificatory feedback is irrelevant to the stability of the HIV classification and the disease it classifies

---

[20] The terminology adopted by Tsou (2007) is confusing since Hacking uses 'looping effects' to refer to cases wherein a human science classification changes classified people, thereby requiring revisions to the classification. What Tsou (2007) calls 'weak implications of looping' is referred to herein as 'classificatory feedback'; what he calls 'strong implications of looping' is referred to as 'looping effects.'

[21] Tekin (2011, 2014) offers a rich elaboration and expansion of classificatory feedback by focusing on the concepts of the self, self-concept, and narratives, which are underdeveloped in Hacking's account.

(although classifications can obviously be revised as scientists gain more knowledge). It is important to notice that projectable inferences yielded by medical classifications such as HIV—including inferences about treatment (e.g., antiretroviral drug treatment)—are grounded in the fact that the classification refers to a stable biological kind and that patients diagnosed with HIV share relevant biological properties. Hence, HPC kinds are stable objects of classification *despite the presence of classificatory feedback*.[22]

4.3 Psychiatric Classifications that Individuate Biological Kinds are Projectable

Some mental disorders are HPC kinds insofar as they are classes of abnormal behavior constituted by sets of biological mechanisms that interact to produce the key features of the kind. In section 2.2, I presented evidence that mental disorders such as schizophrenia and depression are underwritten by biological mechanisms. Schizophrenia is a relatively rare disorder constituted by biological mechanisms (e.g., excessive dopamine activity in the mesolimbic pathway, deficient glutamate activity) that interact to produce the symptoms of schizophrenia. Similarly, depression is a relatively common disorder constituted by biological mechanisms (e.g., deficient monoamine activity, hyperactive neuroendocrine response, disrupted neuroplasticity) that interact to produce the feelings of sadness associated with depression. Insofar as these disorders are underwritten by sets of stable biological mechanisms, they are HPC kinds constituted by a partly intrinsic (biological) essence.

For mental disorders that are HPC kinds, the presence of a partly intrinsic biological essence (i.e., sets of interacting biological mechanisms) ensures their relative stability and that their classifications yield robust projectable inferences. While classificatory feedback will causally effect individuals diagnosed with these disorders, this feedback will not change the biological mechanisms nor the general psychological and behavioral features (e.g., psychosis, feelings of sadness) defining these disorders. Classifications of HPC kinds yield projectable inferences because these generalizations are *causal consequences of the intrinsic biological essence of these disorders*. For example, the projectable inference that psychosis can be treated with dopamine antagonist drugs is a causal consequence of the intrinsic biological mechanisms that underwrite

---

[22] Conversely, some philosophers (e.g., Bogen, 1988; Cooper, 2005, ch. 2; Haslanger, 2012) demonstrate that classifications of non-human kinds (e.g., 'marijuana,' 'boron,' 'food') can change these kinds via classificatory feedback. Khalidi (2010) makes the stronger claim that *looping effects* are not limited to the human domain. For example, the classification of wolves (*Canis lupus*) *as domesticated animals* led—through selective breeding—to the creation of dogs and a new species classification (*Canis familiaris*), which required revisions to the original classification (i.e., the addition of *Canis lupus familiaris* as a subspecies).

psychosis (e.g., excessive dopamine activity in the mesolimbic pathway). Similarly, the projectable inference that the signs of depression can be alleviated with serotonin agonist drugs is a causal consequence of the fact that depression is underwritten by stable biological mechanisms (e.g., deficient serotonin activity). These considerations concerning projectability undercut skeptical arguments against the usefulness of analyzing mental disorders as natural kinds (e.g., see Zachar, 2000, Tekin, 2016).

It is important to note that the projectable inferences yielded by classifications of HPC kinds are *ampliative* insofar as they make predictions that go beyond the nominal criteria that are used to identify kind members. Some commentators obscure this point by suggesting that the inductive inferences that HPC kind terms support are predictions concerning the (surface) properties used to identify members of a kind. For example, Ereshfsky (2017) states that:

> [M]embers of *Canis familiaris* ... tend to share a number of common properties (having four legs, two eyes, and so on)... For [the] HPC theory, the similarities among the members of a kind must be stable enough to allow better than chance prediction about various properties of a kind. Given that we know that Sparky is a dog, we can predict with greater than chance probability that Sparky will have four legs.

Ereshefsky's characterization is misleading insofar as it fails to appreciate the *types* of projectable inferences the HPC account is motivated to explain (cf. Tekin, 2016, p. 149). The inferences that Ereshefsky cites are *non-ampliative* insofar as they do not offer predictions going beyond the surface properties (or 'nominal essence') of such kinds. Significantly, these non-ampliative inferences will be yielded by classifications of *artificial kinds*. For example, one can 'predict' that if the bible is a book, then it will include words; or one can 'predict' that if Sally is a widow, then she will be a woman. The chief epistemic benefit of Boyd's theory of HPC kinds is that it can explain how we can draw ampliative projectable inferences that go beyond the nominal criteria that are used to identify its members. As Boyd (1999a) puts it: "[W]hat the [HPC] theory of natural kinds helps to explain, is how we are able to identify *causally sustained regularities* that go beyond actually available data" (p. 152). Elsewhere, Ereshefsky (2010) criticizes essentialist accounts of HPC kinds, in part, because they offer no theoretical benefits. As argued herein, the chief benefit of (partly) intrinsic essentialist accounts of HPC kinds is that they explain how HPC kind terms yield robust and ampliative projectable inferences. In the case of mental disorders, this importantly includes inferences concerning the prognosis of a disorder and possible treatment interventions. It is these types of ampliative inferences that distinguish HPC kinds from artificial kinds and illuminate their importance in scientific contexts.

4.4 HPC Kinds, Social Mechanisms, and the Expression of Mental Disorders

My analysis of HPC kinds suggests that classifications of mental disorders constituted by (some) intrinsic biological mechanisms will yield robust and ampliative projectable inferences. But what about social mechanisms? One apparent difference between medical diseases and mental disorders—due to the effect of social mechanisms—is that the characteristic signs of mental disorders are subject to much more variability across cultures. Moreover, social and cultural factors undoubtedly play a central role in the emergence of 'transient mental illnesses' (e.g., hysteria, dissociative fugue) that only appear in specific historical periods (Hacking, 1998) and 'culture-bound syndromes' (e.g., *dhat, latah*) that only appear in particular cultures.

Anthropological research supports the inference that social mechanisms can stabilize a unique and specific expression of mental disorders in particular cultural contexts. Cross-cultural research indicates that mental disorders such as schizophrenia, depression, and anxiety disorders appear in all cultures, but the expression of these disorders varies across cultures (Kleinman, 1988, chs. 2–3). To make sense of these findings, I maintain that the similarities reflect common biological mechanisms, whereas differences reflect social and cultural mechanisms. Moreover, the characteristic signs of some mental disorders (e.g., schizophrenia) are expressed *more uniformly* across cultures compared to other disorders (e.g., depression, anxiety). For example, in non-Western societies (e.g., China), depression is more typically expressed as somatic complaints (e.g., headaches, dizziness, lack of energy); conversely, in Western societies, depression is more typically expressed as feelings of guilt (Kleinman, 1988).[23] One way to interpret these findings is to regard the *uniformity of a condition across cultures* as a measure of the extent that a disorder is *determined by biological mechanisms* (Tsou, 2007, 2013). On this view, some mental disorders (e.g., schizophrenia, bipolar disorder) are *more natural* than others because their characteristic signs are *more directly determined by biological mechanisms*. By contrast, disorders whose characteristic signs are more strongly mediated by social mechanisms (e.g., hysteria, bulimia) are *more artificial* (although there may be a shared biological basis for these disorders). This view is consistent with the finding that the symptoms of the most severe and debilitating mental disorders (e.g., schizophrenia, chronic depression) are expressed more uniformly across cultures (Marsella, 1988).

---

[23] Research suggests that somatic-depression is more common globally, while guilt-depression is a cultural variant specific to Western cultures (Kirmayer, 2001).

The framework advanced herein suggests that biological mechanisms determine the general psychological and behavioral features of mental disorders (e.g., psychosis, mania, depressive states), whereas social mechanisms determine a more specific, culturally-sanctioned *expression* of a disorder (e.g., religious delusions, feelings of guilt). Kleinman (1988) criticizes this theoretical framework—what he dubs the 'pathogenetic/ pathoplastic model'—for overemphasizing the influence of biological factors in determining the outcome of a disorder (pp. 24–7). In its place, Kleinman recommends what he calls a 'dialectal model':

> [A] more useful model is one in which biological and cultural processes dialectically interact. At times one may become a more powerful determinant of outcome at other times the other. Most of the time it is the interaction, the relationship, between the two which is more important than either alone (pp. 25–6).

While the pathogenetic/ pathoplastic model regards biological mechanisms (that determine the general causal structure of a disorder) and cultural mechanisms (that determine the content of disorders) as separable, Kleinman's dialectic model maintains that it is the causal interaction between biological and cultural mechanisms (which are regarded as inseparable) that determines the outcome of disorders. One implication of the dialectical model is that different instantiations of disorders (e.g., guilt-depression versus somatic-depression) across cultures should not be regarded as cultural variants of a single biological kind, but as distinct disorders (or kinds) determined by unique biological-cultural causal interactions. While I accept the dialectical assumption that the interaction between biological and social mechanisms can lead to different *outcomes* for disorders observed in different cultures, I accept the pathogenetic/ pathoplastic assumption that cross-cultural variations reflect different expressions of the same ('universal') biological disorder (or biological kind). This latter assumption is supported by the fact that pharmacological drugs (e.g., antidepressants drugs) are efficacious for treating patients in different cultures who exhibit cultural variants of mental disorders such as depression (Lin, Poland, & Anderson, 1995).

The social mechanisms emphasized by Szasz, Scheff, and Hacking (i.e., broadcasting of stereotypes, labeling, role-adoption) play an important role in *stabilizing a specific expression of a mental disorder* (cf. Mallon, 2016). My disagreement with these authors stems from their neglect of biological mechanisms that determine the general psychological and behavioral features of mental disorders that are HPC kinds. In the case of *more artificial* classes (e.g., 'bulimia'), the stereotypical signs that define these classes are determined *more directly* by cultural mechanisms than biological mechanisms.

This framework can account for *transient* mental disorders (Hacking, 1998), such as hysteria. The hysteria classification was most prominent in France during the late 19th century, and it was characterized by symptoms such as convulsions, seizures, feelings of strangulation, fainting, swooning, paralysis of the limbs, and trancelike states (Goldstein 1989, pp. 323–4). In the 1880s, a common stereotype of hysteria (viz., it is a neurological condition, typically afflicting females, and caused by past trauma) was popularized in French culture by various sources (e.g., Charcot's famous lectures at the Salpêtrière where he treated of hysterical patients with hypnosis, depictions of hysteria in photographs). This period was followed a dramatic increase in cases of hysteria, which gradually declined during the 20th century to its virtual disappearance in 1980, when the hysteria classification dissolved and was split into several DSM categories, viz., 'conversion disorder,' 'somatization disorder,' and 'histrionic personality disorder' (Micale, 1993). On the assumption that there was *some* biological basis for the distress expressed by hysteric patients (e.g., the biological mechanisms implicated in anxiety and depression), social mechanisms (e.g., labeling, role adoption) played a crucial role in *stabilizing the specific set of symptoms* (e.g., fainting, swooning, paralysis) *expressed by hysterics*.

The social mechanisms (e.g., role adoption, imitation of stereotypes) that stabilize transient mental disorders also stabilize *culture-bound syndromes* (CBS) that only occur in certain cultures (APA, 2000, p. 898).[24] While some CBS (e.g., *koro, locura*) appear to be culture-specific expressions of psychosis, many CBS (e.g., *susto, latah, ataque de nervios*) appear to be culture-specific expressions of depression, anxiety, or a combination of both. My analysis suggests that CBS are not different in kind from cultural variants of universal disorders (e.g., guilt-depression) or transient mental disorders (e.g., hysteria). With respect to the *naturalness* of CBS, these disorders are natural to the extent that they are constituted by stable biological mechanisms, which provide the basis for robust projectable inferences. By contrast, the specific expression of such disorders that are determined by social mechanisms represent *artificial* (or socially constructed) aspects of CBS.

Rachel Cooper (2010) argues against the view that CBS should be regarded as variants of a more general (or 'universal') disorder, and she suggests that *some* CBS are distinctive natural kinds. In articulating this argument, Cooper (2010) draws an analogy between psychiatry and geology:

---

[24] In DSM-5 (APA, 2013, p. 758), CBS are replaced with three constructs, which are claimed to have greater clinical utility: (1) *cultural syndrome*, (2) *cultural idioms of distress*, and (3) *cultural explanations*.

[I]f . . . different varieties of anxiety/depression type disorders are formed into distinct entities by cultural context, then we can think of such disorders as being kinds analogous to the different kinds of rock distinguished by geologists. In both cases a basic amorphous "material" would . . . take different forms depending on the history of its formation. Kinds that only occur in specific historical contexts—such as kinds of rock, and culturally formed types of mental disorder—can usefully be considered scientific kinds . . . because they can support explanations and inductive inferences and feature in law-like generalisations. . . . Admittedly, the members of such kinds are only found in certain environments. . . . [Although] cases may occur for a limited time or in limited places, within those constraints the kinds are as scientifically useful as those that occur universally. (pp. 330–1)

Cooper's qualifications regarding the limited stability of CBS and limited projectable inferences yielded by their classifications motivates an argument against her conclusion. Tsou (2020) argues that classifications underwritten by biological mechanisms yield relatively stable (or robust) projectable inferences, while classifications underwritten by social mechanisms yield unstable and transitory projectable inferences. Assuming that cultural evolution moves at a much faster pace than biological evolution (Richerson, Boyd, & Heinrich, 2010; Perrault, 2012) and that psychiatry should formulate categories that yield *stable predictions*, psychiatry should classify kinds *at levels* that correspond to biological kinds. This implies CBS should be classified as variants of a more general biological kind (e.g., depression) because classifications that individuate biological kinds yield more stable projectable inferences than classifications formulated at more specific (folk) levels that incorporate the effects of social mechanisms (cf. Blease, 2010). With respect to the social factors (e.g., cultural norms, formation of stereotypes) that determine specific features of CBS, it is worth noting that Cooper's comparison with geology is inapt: whereas the physical environments that give rise to specific rocks remain relatively stable over long evolutionary time-scales, the cultural environments that give rise to CBS change much more rapidly.

4.5 Biological Kinds are Useful Objects of Psychiatric Classification

The account of HPC kinds in psychiatry defended in this section supports the proposal advanced in section 3.5 that mental disorders are biological kinds with harmful effects. In emphasizing the stability of HPC kinds and the robust projectable inferences yielded by their classifications, one of my aims is to illustrate that HPC kinds are appropriate objects of psychiatric classification because these classes are characterized by a set of stable biological mechanisms that researchers can identify. Understanding how biological mechanisms are causally related to abnormal psychological and behavioral signs also provides a means for identifying successful clinical interventions for individuals who display signs of these disorders. What this account rules out as genuine mental disorders are artificial kinds: classes of abnormal behavior that are not associated with any distinctive biological mechanisms.

My proposal to liberalize the naturalistic component of mental disorder away from biological dysfunction (or pathology) to the more deflationary criterion of biological kinds is based on pragmatic considerations concerning treatment. Psychiatry aims to help individuals deal with conditions that are harmful insofar as they impede their capacity to live a normal life. From a pragmatic perspective, the question of whether a condition is the result of a biological dysfunction is irrelevant to whether that condition should be treated or not. On the other hand, the question of whether individuals diagnosed with a mental disorder will respond to interventions in *predictable ways* is relevant. The criterion of being a biological kind provides assurance that classified mental disorders will yield robust and ampliative projectable inferences relevant to prognosis and treatment. In this regard, medical doctors routinely treat conditions that do not involve biological dysfunction. For example, doctors treat conditions due to injury (e.g., a broken leg) or conditions that involve normal physiological responses (e.g., chronic pain). In the context of psychiatric treatment, if it turned out that depression or PTSD reflect normal (and law-like) *biological and psychological responses* to experiencing deeply traumatic events, ruling out studying or treating such conditions because they do not involve biological dysfunction would appear to be pragmatically indefensible. These pragmatic considerations motivate my argument in section 5.3 that biological kinds are appropriate targets of psychiatric classification.[25]

5 Psychiatric Classification and the Pursuit of Diagnostic Validity

The *Diagnostic and Statistical Manual of Mental Disorders* (DSM), published by the American Psychiatric Association (APA), is the most authoritative psychiatric classification manual. Historically, the third edition of the DSM (DSM-III) was decisive in establishing its scientific credibility. DSM-III introduced its purely descriptive approach to classification, whereby mental disorders are defined by a set of necessary and sufficient 'diagnostic criteria.' While DSM-III improved the (interrater) reliability of its

---

[25] These considerations demonstrate that the contrast Zachar (2000, 2014, ch. 9) draws between 'natural kinds' and 'practical kinds' is not as stark as he presents. The account of HPC kinds defended herein offers a naturalistic account that can *explain* the 'stable patterns' or 'pragmatic successes' sought by the practical kinds model.

categories, the DSM has failed to provide *valid* diagnostic categories that *accurately represent real classes*. This is largely due to its theorization of mental disorders as *discrete disease-syndromes*. I argue that the DSM should reconceptualize its objects of classification as biological kinds. I subsequently examine Tabb's argument, which draws on methodological assumptions of the *Research Domain Criteria* (RDoC): the DSM has failed to formulate valid categories because its definitions generate heterogenous groups of patients unsuitable for discovering biomedical facts. Tabb's assessment neglects a more basic problem with the DSM: its failure to revise its categories to incorporate scientific findings. To address this problem, the DSM should replace its purely descriptive methodology with a causal methodology, which would provide a more transparent and testable approach for formulating valid diagnostic categories.

5.1 Concepts of Diagnostic Validity

While there is no agreed upon concept of validity in psychology or psychiatry, valid diagnostic categories are generally understood as definitions of mental disorders that 'carve nature at the joints' in the sense of *corresponding to real classes in nature* (Jablensky, 2016). Concepts of *diagnostic validity* draw upon psychometric concepts developed for psychological tests. In psychometrics, validity is a measure of a test's *accuracy*. A psychological test is valid to the extent that it measures what it purports to measure (e.g., 'intelligence'): "the degree to which empirical evidence and theoretical rationales support the adequacy and appropriateness of interpretations and actions on the basis of test scores" (Messick, 1995, p. 741). Applying psychometric concepts of validity to the DSM's diagnostic categories yields the following concepts of diagnostic validity:

(1) *Construct validity*: operationalizations of a diagnostic category (e.g., 'major depressive disorder') accurately represent a theoretical construct (i.e., depression as a discrete disease-entity).

   a. Convergent validity: a category correlates with independent measures (e.g., depression scales and tests) that purport to measure the same construct.

   b. Discriminant validity: a category does not correlate with different constructs (e.g., 'generalized anxiety disorder') assumed to be distinct and unrelated.

(2) *Content validity*: the criteria used to define a diagnostic category are an adequate and representative sample of the criteria relevant to the construct.

(3) *Criterion-related validity*: a diagnostic category is correlated with external criteria (e.g., treatment response) that are assumed to be representative of the construct.

   a. Concurrent validity: a category correlates with a criterion (e.g., biomarkers) at the time of diagnosis.

   b. Predictive validity: a category can predict criteria after the diagnosis (e.g., prognosis).

Construct validity is the most important and overarching type of validity. Construct validation is a complex process that requires evidence from multiple independent sources (including content and criterion validators) that converge on the inference that an operational definition accurately represents the theoretical construct it intends to represent (Cronbach & Meehl, 1955). Hence, construct validity is more general than (and a prerequisite for establishing) content and criterion-related validity insofar as the former is required (and sets the upper limit) for the latter two (Nelson-Gray, 1991).

Analyses of diagnostic validity by biological psychiatrists focus on construct and criterion-related validators. In their foundational paper, Robins and Guze (1970) introduced the term 'diagnostic validity' and articulated five phases that facilitate the validation of diagnostic categories (cf. Feighner et al., 1972): (1) clinical description of the disorder, (2) incorporation of findings from laboratory (e.g., chemical, physiological, anatomical) studies, (3) delimitation of a diagnostic category from other categories with exclusionary criteria, (4) follow-up studies to confirm patients were correctly diagnosed, and (5) family studies to assess the history of the disorder in the patient's family. Robins and Guze emphasize that these five phases interact with one another and constitute an iterative process of revision and refinement directed towards *more homogenous diagnostic groups*.[26] Kendler (1980) expanded on this analysis by distinguishing three criterion-related validators: (1) antecedent (or 'postdictive') validators (e.g., familial aggregation, premorbid personality), (2) concurrent validators (e.g., psychological tests), and (3) predictive validators (e.g., response to treatment, rates of recovery). Andreason (1995) updated the kinds of laboratory studies considered by Robins and Guze, arguing that the results of neurobiological studies (e.g., molecular genetics, neurochemistry, neurophysiology) provide additional criterion-related validators that could *causally link the criteria used to define diagnostic categories with neural substrates*. Similarly, Hyman (2007) argued that DSM categories could be validated by the results of neurobiological studies, such as studies on genetic risk factors and neuroimaging studies that demonstrate structural and functional abnormalities.

---

[26] Although Robins and Guze never use the term, these five phases are different methods for achieving construct validity (Schaffner, 2012).

The criterion-related validators (i.e., biomarkers) emphasized by Andreason and Hyman should be regarded as a gold standard of diagnostic validators insofar as they provide compelling evidence that a disorder has a *distinctive biological basis*. This captures the idea that the least controversial mental disorders (e.g., Down syndrome, phenylketonuria, Alzheimer's disease, Huntington's disease) are associated with clear and distinctive biomarkers (Kendell & Jablensky, 2003).

5.2 Most DSM Categories are Invalid

There is scarce evidence that any DSM diagnostic categories—other than a small handful (viz., 'schizophrenia,' 'bipolar disorder,' 'intellectual disability,' 'neurocognitive disorders')—possess construct validity. To have construct validity, a diagnostic category should accurately represent a construct *as defined by theory*. While DSM-III was *presented* as an "atheoretical" manual that makes no assumptions about the causes of mental disorders (APA, 1980, p. 7), its architects advocated the biological ('neo-Kraepelinian') approach to psychiatry championed by Eli Robins and Samuel Guze, which assumed that mental disorders are *discrete diseases* (Blashfield, 1984; Wilson, 1993).[27] This theoretical assumption is reflected in the formal definitions of mental disorder found in various editions of the DSM (e.g., see APA, 1980, p. 6; APA, 2000, p. xxxi). In DSM-5, "mental disorder" is defined as: "[A] syndrome characterized by clinically significant disturbance in an individual's cognition, emotion regulation, or behavior that reflects a dysfunction in the psychological, biological, or developmental processes underlying mental functioning" (APA, 2013, p. 20).

The DSM's explicit definitions of mental disorder imply that its diagnostic categories possess construct validity to the extent they individuate discrete syndromes caused by dysfunctional (i.e., biological, psychological, or developmental) processes. Hence, there should be evidence that a diagnostic category individuates: (1) a discrete condition with natural boundaries ('zones of rarity') that distinguish it from other disorders and normality, and (2) a condition caused by dysfunctional processes (Kendell & Jablensky, 2003). Most DSM categories fail to meet both of these criteria. Regarding the discreteness assumption, which is associated with the categorical (as opposed to dimensional) approach to psychiatric classification championed in DSM-III, Kendell and Jablensky (2003) point out that that when Robins and Guze

---

[27] This assumption was stated explicitly in publications by members of the DSM-III taskforce (e.g., Robert Spitzer, Jean Endicott) prior to the publication of DSM-III (Tsou, 2011, p. 464). The DSM-III taskforce planned to include the definition "mental disorders are a subset of medical disorders" in DSM-III; however, this statement was ultimately not included (Mayes & Horwitz, 2005).

(1970) published their classic analysis, it was widely assumed that disorders such as schizophrenia and bipolar disorder were transmitted by distinct sets of genes. As discussed in section 2.2, this assumption is false. The largest genome-wide association studies (GWAS) suggest a common genetic vulnerability for schizophrenia and bipolar disorder. Others argue that the discreteness assumption is violated by the high co-occurrence ('co-morbidity') of allegedly distinct disorders, e.g., depression and anxiety disorders (Sullivan & Kendler, 1998; Hyman, 2007; Aragona, 2009). Regarding the dysfunction assumption, while there is plausible evidence that some disorders are caused by biological dysfunction (e.g., schizophrenia, bipolar disorder, autism),[28] other disorders (e.g., anxiety, PTSD) are arguably *normal psychological reactions* to stress and trauma (Lilienfeld & Marino, 1995). Depression is a case in point. Horwitz and Wakefield (2007) argue that many conditions that are currently diagnosed as major depression are *normal psychological responses to life circumstances*, rather than pathological conditions caused by dysfunctional processes. While these authors maintain that there are 'genuine' cases of depression caused by dysfunctional processes, the category of 'major depressive disorder' is specified too broadly to distinguish dysfunctional cases from normal cases of sadness. Most of the fine-grained diagnostic categories in the DSM (e.g., 'narcissistic personality disorder,' 'binge eating disorder,' 'separation anxiety disorder') lack construct validity insofar as there is little evidence that these categories correspond to discrete diseases caused by dysfunctional processes.

Some diagnostic categories of the DSM, including consensus disorders, lack content validity insofar as the diagnostic criteria that define disorders fail to adequately represent the category. As discussed in section 2.2 (see note 11), the exclusion of anxiety in the diagnostic criteria for depression is inconsistent with research that indicates: (1) anxiety is a core sign of depression that appears among 75% of depressed patients, and (2) depression and anxiety co-occur around 60% of the time. The exclusion of anxiety from the diagnostic criteria for depression appears to be a decision made by the DSM-III taskforce based on an ideological commitment to the classificatory principle that mental disorders are *discrete* diseases, rather than empirical evidence. This interpretation is

---

[28] The high heritability of schizophrenia, bipolar disorder, and autism (with heritability estimates greater than 80%) distinguishes them as plausible candidates for mental disorders caused by biological dysfunction. The heritability of these disorders is much higher than the heritability estimates of consensus diseases, such as breast cancer (5–60%) and Parkinson's disease (13–30%), for which genetic risk factors have been identified (Burmeister, McInnis, & Zöllner, 2008). Evidence that these disorders are caused by biological dysfunction would require the identification of genetic risk factors and disrupted functional mechanisms. For genetic risk factors and disrupted mechanisms in schizophrenia, see section 2.2.

supported by Lee Robins' retrospective report that a principle used in the construction of DSM-III was that "the same symptom could not appear in more than one disorder," which she regarded as a poor rule (cited in Maj, 2005, p. 182). From the perspective of general medicine, Robins' assessment seems to be correct since there are different illnesses (e.g., viral illness, bacterial illness) that cause similar symptoms (e.g., flu-like symptoms).[29]

While many DSM diagnostic categories lack criterion-related validity (i.e., correlation with external criteria), this issue requires careful assessment given that criterion-related validity (especially predictive validity) is one of the most concrete and important types of diagnostic validity. Kendell and Jablenksy (2003) argue that most diagnostic categories of the DSM lack validity, but some categories (e.g., 'schizophrenia,' 'bipolar disorder') possess *clinical utility* insofar as they provide "nontrivial information about prognosis and likely treatment outcomes, and/or testable propositions about biological and social correlates" (p. 9). This distinction is peculiar since what Kendell and Jablensky call 'utility' is precisely what has traditionally been called predictive validity (First et al., 2004; Schaffner, 2012). In this light, some diagnostic categories in the DSM possess *some* predictive validity insofar as they yield reliable predictions about prognosis, possible treatments, and biomarkers.[30] For example, individuals exhibiting the signs of schizophrenia (e.g., psychosis) are likely to benefit from dopamine antagonist drug treatment, or individuals displaying the signs of bipolar disorder (e.g., mania) are likely to benefit from lithium treatment. In this connection, while broad categorical distinctions between schizophrenia, bipolar disorder, and depression may yield reliable predictions relevant for treatment, the kinds of fine-grained diagnostic categories found in the DSM (e.g., 'histrionic personality disorder,' 'disruptive mood dysregulation disorder,' 'intermittent explosive disorder') are less useful because they fail to yield predictions *specific to these categories* (Tsou, 2015).

---

[29] In this connection, cognitive impairments (e.g., disturbances in executive function) are common in schizophrenia and depression. It is well established that cognitive impairments are a core symptom of schizophrenia that should be included in the DSM criteria for schizophrenia (Hyman, 2010). Arguably, cognitive impairments should also be included in the DSM criteria for depression given their presence in severe forms of depression (Tsou, 2013). Similar mechanisms (e.g., disrupted neuoroplasticity, functional abnormalities in the dorsolateral PFC) are implicated in the cognitive impairments observed in schizophrenia and depressions (Lewis & González-Burgos, 2008; Pittenger & Duman, 2008).

[30] I reject the assumption that validity is an all-or-nothing matter, whereas utility allows for gradients (Kendell & Jablensky, 2003, p. 10). If some categories are *more* predictively valid than others (e.g., produce more reliable projectable inferences), then predictive validity is indistinguishable from Kendell and Jablensky's definition of utility.

5.3 The DSM Should Classify Biological Kinds, Not Diseases

The DSM has failed to provide categories with construct validity because it has theorized mental disorders narrowly as *medical conditions* ('syndromes') caused by dysfunctional processes. Against this view, I argue that the DSM should theorize mental disorders more broadly to be natural kinds (i.e., classes of abnormal behavior underwritten by biological mechanisms) that are harmful.

The main reason why the DSM has failed to formulate diagnostic categories with construct validity is due to its vague and minimal theorization of mental disorders as disease syndromes: classes of abnormal psychological and behavioral symptoms that are caused by dysfunctional (biological, developmental, and psychological) processes. Given the desire to unify psychiatry with medicine, it is understandable that the DSM's definitions of mental disorder implicitly presuppose a medical model that conceptualizes abnormal behavior as 'symptoms' of underlying 'pathological' processes (Guze, 1993). However, empirical evidence does not support the inference that many of the diagnostic categories of the DSM, including consensus disorders (e.g., depression, anxiety) are caused by *dysfunctional* processes. As discussed in section 3.5, judgments regarding the presence of dysfunction are difficult to ascertain, even when trying to apply precise accounts of biological dysfunction, such as those articulated by Boorse and Wakefield. Assessments of dysfunction are even more difficult to discern when adopting the DSM's more ambiguous appeal to dysfunctional *psychological, biological, or developmental processes*. Yet, it is obvious that there is not compelling evidence that many DSM categories (e.g., 'dependent personality disorder,' 'oppositional defiant disorder,' 'voyeuristic disorder') result from dysfunctional processes. At the very least, the hypothesized dysfunctional processes involved in these mental disorders have not been clearly articulated nor received widespread consensus.

As argued in section 4.5, the DSM should classify biological kinds, rather than diseases. The argument that the targets of psychiatric classification (i.e., mental disorders) should be biological kinds is supported by a number of pragmatic and scientific considerations. First, the understanding of mental disorders as biological kinds offers a transparent criterion for determining when a diagnostic category is valid and should be included in the DSM: mental disorders should be associated with a *set of biological mechanisms that cause a characteristic set of abnormal behaviors*. In ideal cases (e.g., schizophrenia), there should be evidence that the criteria used to define a diagnostic category (e.g., negative symptoms) are the result of distinct biological mechanisms (e.g., deficient dopamine activity in the mesocortical pathway). Second, the requirement that diagnostic categories individuate biological kinds ensures that categories yield robust and ampliative projectable inferences (e.g., predictions related to prognosis or treatment).[31] Finally, the theorization of mental disorders as biological kinds can accommodate the different types of conditions that are currently categorized as mental disorders. While some biological kinds are diseases caused by dysfunctional biological processes (e.g., schizophrenia, bipolar disorder), other biological kinds are psychological reactions (e.g., acute depression) caused by normal biological processes.[32]

5.4 The RDoC: A Psychiatric Classification System for Research

Disenchantment with the DSM spurred the development of an alternative psychiatric classification system in the 2010s by the National Institute of Mental Health (NIMH): The *Research Domain Criteria* (RDoC). The RDoC was formulated due to dissatisfaction with the DSM's failure to provide valid diagnostic categories that could effectively guide research. DSM categories are an impediment to research because they are utilized by default to compile populations of patients to study. Without prior assurance that DSM categories are valid, the reliance on DSM categories in research contexts could *reify* psychiatric constructs that have no real (i.e., biological) basis (Hyman, 2010). This practice falls well short of the process of diagnostic validation envisioned by Robins and Guze (1970), wherein diagnostic categories individuate increasingly *homogenous* groups over time.

In contrast to the DSM, the RDoC is a classification system formulated exclusively for research purposes and informed by biological and behavioral science (Faucher & Goyer, 2015; Zachar & Kendler, 2017). While the DSM classifies psychiatric constructs at the level of mental disorders (or symptom-clusters), the RDoC classifies constructs (e.g., *loss, potential threat, circadian rhythms*) at lower levels of analysis. This shift in targets of psychiatric classification reflects the RDoC's methodological assumption that a bottom-up approach to validating psychiatric constructs is more promising than the DSM's top-down approach of prescribing operational definitions of mental disorders that can *subsequently* be validated by research. In this manner, the psychiatric constructs specified by the RDoC offer more *direct* targets of validation than the DSM. In championing a more concrete and empirically-

---

[31] This perspective is amenable to the 'spectrum disorders' approach (Hyman, 2007), which assumes that a diverse range of symptom-profiles can be caused by the same mechanisms. If the DSM classified biological kinds, some fine-grained categories (e.g., phobias, agoraphobia) could be reclassified as sub-types of a more general spectrum category (e.g., fear and avoidance disorders), provided they are underwritten by similar biological mechanisms (Kupfer & Regier, 2011).

[32] Normal psychological reactions (e.g., depression, anxiety) are suitable candidates for disorders that could be classified dimensionally (e.g., utilizing dimensional measures to indicate different severity levels), rather than categorically (Hyman, 2007; Kupfer, First, & Regier, 2002).

driven approach to psychiatric validation, the RDoC aims to connect narrowly-defined psychiatric constructs (e.g., *loss, perception, frustrative nonreward*) to biological correlates. The RDoC distinguishes its objects of classification ("constructs") in terms of different "domains" and "units of analysis," which is encapsulated in the RDoC matrix (see figure 1):

The constructs classified by the RDoC are categorized under one of the general domains and are regarded as hypothesized concepts "regarding brain organization and functioning" (Morris & Cuthert, 2012, p. 30). For example, the domain of 'negative valence systems' includes constructs such as *active threat ('fear')*, *potential threat ('anxiety')*, and *loss*; the domain of cognitive systems includes constructs such as *attention, working memory*, and *cognitive (effortful) control*. As implied by the domains (e.g., 'positive valence systems') specified by the RDoC, there is no assumption that domains or constructs necessarily involve pathological or dysfunctional processes (Cuthbert & Insel, 2010, p. 313). Moreover, there is no assumption that the constructs involved in mental disorders only come from a single domain; multiple domains may be involved in and cross-cut different disorders (cf. Khalidi, 1998; Hoffman & Zachar, 2017). RDoC constructs are assumed to correlate with measures for at least one of the units (or levels) of analysis (e.g., genes, neural circuits). Since the first version of the RDoC matrix was published in 2016, other constructs and subconstructs have been added to the matrix, and an impressive array of biological correlates ('units of analysis') have been causally linked to these constructs (see NIMH, 2018).

5.5 What are Appropriate Targets of Psychiatric Classification?

Drawing on methodological assumptions of the RDoC, Kathryn Tabb (2015) rejects the DSM's assumption of *diagnostic discrimination*, which maintains that "the operationalized criteria for diagnosing clinical types will … successfully pick out populations about which relevant biomedical facts can be discovered" (p. 1048). Tabb rejects this assumption on historical and methodological grounds. From a historical perspective, Tabb argues that the historical origins and trajectory of the DSM provide no reason to be optimistic for vindicating this assumption. From a methodological perspective, she argues that the *heterogeneous* groups of patients produced by DSM definitions fail to yield suitable research samples for discovering underlying mechanisms. This argument emphasizes the polythetic nature of DSM categories (First & Westen, 2007) insofar as the diagnostic criteria for mental disorders generate an immense array of dissimilar symptom-profiles. Tabb praises the RDoC for providing an alternative way of assembling research samples and shifting research away from DSM definitions. Citing Hyman and Fenton (2003) approvingly, Tabb (2015) writes:

RDoC Matrix								
	Units of Analysis							
Domains	Genes	Molecules	Cells	Circuits	Physiology	Behavior	Self-Report	Paradigms
Negative Valence Systems								
Positive Valence Systems								
Cognitive Systems								
Systems for Social Processes								
Arousal/ Regulatory Systems								

Figure 1 The RDoC Matrix

(adapted from Morris and Cuthbert, 2012)

> By encouraging the funding of research that investigates certain research
> domains at certain units of analysis, RDoC changes the targets of validation
> from "clinical endpoints that have remained unchanged for decades" to any
> "sort of phenomenon relevant to psychopathology that may be viewed either
> as an extreme on a spectrum of human variation or as a dysfunctional
> structure or process." (p. 1052)

Tabb opposes the overreliance on DSM categories for compiling populations to
study and financial incentives that reinforce their utilization in research contexts, which function to preserve the status quo and prevent investigation of
different psychiatric constructs.

   Tabb (2019) supplements her rejection of diagnostic discrimination with an
argument that philosophers of psychiatry should not focus exclusively on *diagnostic
kinds*. Diagnostic kinds are the traditional categories of mental disorders (e.g.,
'schizophrenia,' 'depression') codified by the DSM, i.e., mental disorders that are
assumed to be diagnosable in both individuals and populations. Tabb complains that
philosophers of psychiatry (e.g., Murphy, 2006; Cooper, 2013; Tsou, 2016) have
assumed that diagnostic kinds are the most important psychiatric kinds to investi-
gate, while ignoring alternative psychiatric kinds. With the emergence of alternative
psychiatric classifications systems that are formulated without reference to DSM
categories (e.g., see Borsboom & Cramer, 2013; Kotov et al., 2017), Tabb argues
that biomedical researchers have moved away from the 'diagnostic kind model,'
wherein the chief goal is to validate diagnostic kinds. These alternative classification
systems promote investigation of different psychiatric kinds (formulated at various
levels of analysis) and reject central assumptions of the DSM (e.g., mental disorders

are discrete rather than dimensional, the biological mechanisms of one disorder cannot cross-cut other disorders). From this perspective, Tabb (2019) argues that: "[P]hilosophers need to either counter psychiatrists' growing suspicion about the hegemony of diagnostic categories in the clinic and the laboratory, or join in redirecting their efforts toward the development of robust accounts of other sorts of psychiatric objects and processes" (p. 2177). Building on the pluralistic approach to classification articulated by Haslam (2003), Tabb opts for the second disjunct, defending a pluralistic view that recognizes and encourages investigation of a multiplicity of psychiatric kinds (e.g., risk factors, personality dimensions, phenomenological states, social conditions) other than diagnostic kinds. In Tabb's vision of integrative pluralism (cf. Mitchell, 2003, 2008; Sullivan, 2014, 2017), philosophers of psychiatry ought to focus their attention on a broader range of psychiatric kinds, which through integrative efforts will provide a more promising path towards obtaining valid and pragmatically useful psychiatric categories.[33]

Tabb convincingly argues that DSM categories have been an impediment in research contexts and philosophers ought to pay more attention to alternative psychiatric kinds. However, she does not demonstrate that the diagnostic kind model (or a more nuanced version of this model) is *incompatible* with the integrative pluralism that she favors. This response echoes an issue that arose during the development of the RDoC regarding the relationship between the DSM and RDoC (Tsou, 2015). While the RDoC was initially presented as a *competing* classification system to the DSM (Insel et al., 2010), the NIMH and APA subsequently issued a statement indicating that the RDoC and DSM are *complementary* systems and that findings from the RDoC could be used to revise DSM categories (Insel & Lieberman, 2013). To the extent that the psychiatric kinds investigated in alternative classification systems can be incorporated and integrated into DSM categories, one of the most valuable fruits of investigating a plurality of psychiatric kinds would be to criticize and stimulate revisions of diagnostic kinds (Tsou, 2015; cf. Bluhm, 2017). Hence, the validation of diagnostic kinds and investigation of a plurality of psychiatric kinds can be complementary rather than mutually exclusive projects.

Tabb does not establish that the diagnostic kind model is the main culprit in impeding the formulation of valid diagnostic kinds. Arguably, a more fundamental problem is the failure of the DSM to revise its diagnostic categories to reflect

---

[33] Tabb (2019) suggests that the concept of *scientific repertoires*—i.e., an ensemble of material and social conditions that allow communities of scientists to engage in multidisciplinary collaborative work and articulate shared goals (Ankeny & Leonelli, 2016)—provides a useful resource for identifying different types of psychiatric kinds.

empirical findings (Murphy, 2006, ch. 9; Hyman, 2002; Insel et al., 2010; Sanislow et al., 2010; Tsou, 2015), which is one of the DSM's intended roles. In terms of the validators discussed by Robins and Guze (1970), not only have DSM categories failed to adequately incorporate laboratory findings, but many DSM categories (e.g., depression) have failed to meet the more basic goal of providing accurate clinical descriptions (i.e., categories with content validity).[34]

Drawing on Carl Hempel's classic argument that a good psychiatric taxonomic system should be informed by theory (Hempel, 1965, ch. 6), Tsou (2011, 2015, 2019) argues that a theoretical and etiological approach to classification would be superior to the DSM's purely descriptive ('neo-Kraepelinian') approach for formulating valid diagnostic categories (cf. Murphy, 2006). Consider the diagnostic category of schizophrenia. As argued in sections 2.2 and 5.2, the descriptive criteria used to define schizophrenia (e.g., the distinction between positive and negative symptoms) possess *some* validity insofar as there is evidence that these symptoms are associated with *distinctive* neurobiological biomarkers (e.g., positive symptoms being associated with excessive dopamine activity). While the DSM definition of schizophrenia has problems (see note 29), it has been *sufficiently valid* to facilitate the discovery of other neurobiological mechanisms (e.g., deficient glutamate activity). This counters Tabb's pessimistic conclusion that diagnostic kinds *cannot* "successfully pick out populations about which relevant biomedical facts can be discovered" (p. 1048).

The schizophrenia classification is instructive on what is desired in a diagnostic kind. First, despite the DSM's official atheoretical stance, the schizophrenia definition adopted since DSM-IV (APA, 1994) is clearly informed by the theoretical distinction (from biological psychiatry) between positive and negative symptoms (Crow, 1980; Andreason & Olson, 1982). The descriptive criteria used to define DSM categories ought to *incorporate theoretical information regarding the causes of characteristic signs*, which offers a more transparent and testable approach to classification (Tsou, 2015). Second, given the way that the schizophrenia category is defined, it *functions* as a monothetic category that specifies necessary and sufficient criteria in terms of a single symptom (viz., psychosis).[35] By contrast, most diagnostic categories in

---

[34] The inability of the DSM to effectively revise its categories is due to numerous factors, including the insular nature of the DSM revision process (Frances & Widiger, 2012; Tsou, 2015; Bueter, 2019b) and the DSM's conservativism with respect to change given its widespread usage in institutional contexts (Kendler & First, 2010; Sadler, 2013).

[35] The DSM's definition of schizophrenia is polythetic, but since psychosis is distinguished into two separate criteria (i.e., delusions and hallucinations), and an individual must (arbitrarily) meet two of the listed criteria, an individual who experienced a psychotic episode would meet the criteria for a schizophrenia diagnosis. Hence, the schizophrenia category *functions* as a monothetic category.

the DSM are polythetic insofar as they indicate a number of characteristic signs without specifying which of these signs are necessary to receive a diagnosis (Schwartz & Wiggins, 1987). Some DSM categories (e.g., 'bipolar I disorder,' 'panic disorder') that possess some validity are similar to the schizophrenia classification insofar as individuals can receive a diagnosis by the presence a single or few characteristic signs (e.g., manic episodes, panic attacks). If my assessment of the (partial) validity of these diagnostic categories is accurate, then monothetic categories may be a more promising way of formulating diagnostic kinds than the polythetic categories one currently finds in the DSM (Parnas, 2015).

Tabb's argument against diagnostic discrimination highlights the invasive institutionalized role that DSM categories play in research (and treatment) contexts, rather than inherent problems with diagnostic kind model. This is related to the ambitious and all-encompassing manner that the DSM has formulated its goals (Tsou, 2015). In one of the clearest statements of its goals, the authors of DSM-IV-TR state that the purpose of the DSM is to simultaneously: (1) *guide treatment* by providing categories that aid clinicians' judgments, (2) *facilitate research* by providing standardized definitions that can be utilized in the study of mental disorders, and (3) *improve communication* among mental health professionals presupposing disparate theoretical assumptions (APA, 2000, pp. xxiii, xxxvii; cf. APA, 2013, p. xli). The DSM fails to effectively meet any of these goals, including its 'highest priority' of guiding treatment decisions; however, the most promising benefit of the DSM is its capacity for improving communication (Tsou, 2015, 2019).

The DSM could minimize its intrusive role in research and treatment contexts by reframing its goals more narrowly to be an *epistemic hub* that aims to improve communication (Kutschenko, 2011). On this ideal, the DSM's definitions would function as *standardized reference points* (or hubs) that can mediate and coordinate among more specific definitions of mental disorders that are utilized in different (e.g., research, treatment) contexts. In this deflationary role, epistemic hubs facilitate the exchange of information and integration of explanations among mental health professionals who endorse incommensurable theoretical assumptions and practical goals. Because epistemic hubs aim primarily to facilitate communication, its description of mental disorders should be formulated in a sufficiently broad manner so they can be connected with more specific descriptions (Kutchensko, 2011, p. 586; cf. Tabb & Schaffner, 2017).[36]

---

[36] The DSM is currently not an ideal epistemic hub because its fine-grained diagnostic categories are defined at a level of precision unsuitable for accommodating incommensurable definitions of mental disorder (Kutschenko, 2011, p. 595). Hyman (2007) points out that the DSM's arbitrary decision to 'split' (rather than 'lump') disorders contributed to its failure to formulate valid

If the DSM reframed its goals to be an epistemic hub, it would play a far less invasive role in treatment and research contexts. Unfortunately, while the DSM has been impoverished in its capacity to guide treatment and facilitate research, it has been very successful in imposing its utilization in treatment and research contexts, and the APA is unlikely to willingly give up its institutional influence in these domains (Hacking, 2013).

5.6 The DSM, Pluralism, and Theoretical Transparency

Diagnostic kinds formulated at the level of mental disorders and psychiatric constructs formulated at alternative levels are appropriate targets of validation. Tabb's call for the investigation of a plurality of psychiatric kinds offers a remedy to the focus traditionally placed on diagnostic kinds by philosophers of psychiatry. The investigation of a plurality of psychiatric kinds may be one of the most valuable resources for criticizing and revising current DSM categories (Tsou, 2015, Bueter, 2019a; cf. Solomon, 2015). On the other hand, the dogmatic entrenchment of some DSM categories and their resistance to revision indicates that the DSM needs to institute stronger measures and more transparent criteria for determining when a diagnostic category requires revision (or removal) from the manual. The inclusion of some dubious diagnostic categories (e.g., 'histrionic personality disorder,' 'voyeuristic disorder') lends credence to the trite criticism that the DSM merely pathologizes social deviance. The presence of such categories in the DSM should be a source of embarrassment for the APA.

The pursuit of valid diagnostic categories would be better served by a theoretical and etiological approach to classification based on biological psychiatry. In this approach, the descriptive criteria used to define diagnostic categories are theorized to be associated with biological causes (Tsou, 2015). This theoretical approach would be superior to the DSM's antiquated neo-Kraepelinian approach because it would provide a more transparent criterion for classifying mental disorders (i.e., biological kinds) and offer a much more readily testable system.[37] While explicitly stating that biological psychiatry should be the theoretical basis for the DSM will be aversive to some, this approach would not exclude relevant scientific research from other fields (e.g., cross-cultural research would be crucial for identifying the core signs of disorders, psychological research would be important for distinguishing

---

categories. My analysis suggests that DSM categories should be individuated at a sufficiently general level to individuate *distinctive biological kinds* (cf. Craver, 2009, pp. 581–2).

[37] The architects of DSM-5 (APA, 2013) were candid about their desire to move the DSM away from its neo-Kraepelinian, purely descriptive origins towards a theoretical and etiological system of classification based on biological psychiatry (e.g., see Kupfer et al., 2002; Regier et al., 2009; Kupfer & Regier, 2011).

relevant psychological mechanisms). Moreover, since the publication of DSM-III, the DSM has aspired to be a manual informed by biological psychiatry, as evidenced by the biological concepts implicit in the schizophrenia definition. It is time for the DSM to be forthright about its theoretical assumptions, which would be a fruitful step towards the formulation of valid diagnostic categories.

6 Conclusion

In this Element, I defended a number of realist and naturalistic positions on philosophy of psychiatry issues. The most central argument advanced was that *genuine mental disorders are biological kinds with harmful consequences* (sections 3.5, 4.5, and 5.3).

In section 2, I provided a defense of biological approaches to psychiatry. Against skeptics (viz., Szasz, Laing, Scheff) who argue that mental disorders are best explained by social mechanisms (e.g., labeling, imitation of stereotypes, role adoption), I argued that mental disorders such as schizophrenia and depression are constituted by identifiable biological mechanisms.

In section 3, I sought to clarify how we should understand mental disorders through a critical examination of Boorse's and Wakefield's definitions. Against these authors, who identify the naturalistic basis of mental disorders with *biological dysfunction*, I proposed that a more useful naturalistic criterion is the requirement that mental disorders are natural kinds. This deflationary position implies that the proper objects of study and treatment in psychiatry are biological kinds with harmful effects. This account includes mental disorders caused by biological dysfunction (e.g., schizophrenia, bipolar disorder), but also includes harmful disorders (e.g., acute depression) caused by normal biological processes.

In section 4, I elaborated this account by responding to a challenge articulated by Hacking: the kinds studied in psychiatry (e.g., mental disorders) are 'human kinds' that are fundamentally different from 'natural kinds.' I rejected Hacking's argument, showing that some mental disorders are *stable* HPC kinds underwritten by intrinsic biological mechanisms. The requirement that HPC kinds have 'partly intrinsic biological essences' (i.e., biological mechanisms shared by its members) explains how their classifications yield robust and ampliative projectable inferences. In examining the role of social mechanisms emphasized by Hacking, Szasz, and Scheff, I articulated a framework wherein the biological mechanisms underwriting mental disorders determine the general structural features of disorders (e.g., psychosis, depressive states), while social mechanisms determine a more specific expression of disorders in different contexts. This framework explains how biological and social mechanisms interact and can be integrated.

In section 5, I argued that the DSM should classify biological kinds rather than diseases. The DSM has failed to provide diagnostic categories with construct validity (i.e., definitions that accurately correspond to real classes as defined by theory) due to its theoretical assumption that mental disorders are discrete *disease-syndromes caused by dysfunctional processes*. While some DSM categories (e.g., 'schizophrenia,' 'bipolar disorder') are valid in this sense, most diagnostic categories are not. Biological kinds are more useful targets of classification because they offer a more transparent theoretical criterion for construct validity and their classifications yield ampliative projectable inferences (e.g., predictions about treatment). I subsequently examined the methodological assumptions of the RDoC and Tabb's argument that the detrimental role of DSM categories in guiding research should make philosophers wary about the project of validating of diagnostic categories. While Tabb's argument has merit, the inability of the DSM to formulate valid categories stems from its more fundamental failure to revise its categories to incorporate empirical findings. I concluded that the DSM should adopt a theoretical (i.e., biological) and etiological approach to classification and reframe its goals more narrowly to be a standardized reference manual or 'epistemic hub' that aims to facilitate communication and accommodate of a plurality of theoretical and taxonomic perspectives.

References

Amundson, Ron, & Lauder, George V. (1994). Function without purpose: The uses of causal role function in evolutionary biology. *Biology & Philosophy*, 9(4), 443–69.

Andreasen, Nancy C. (1995). The validation of psychiatric diagnosis: New models and approaches. *American Journal of Psychiatry*, 152(2), 161–2.

Andreasen, Nancy C., & Olsen, Scott A. (1982). Negative v positive schizophrenia: Definition and validation. *Archives of General Psychiatry*, 39(7), 789–94.

Ankeny, Rachel A., & Leonelli, Sabina (2016). Repertoires: A post-Kuhnian perspective on scientific change and collaborative research. *Studies in History and Philosophy of Science*, 60, 18–28.

APA (1980). *Diagnostic and Statistical Manual of Mental Disorders, 3rd ed.: DSM-III*. Washington, DC: American Psychiatric Association.

APA (1994). *Diagnostic and Statistical Manual of Mental Disorders, 4th ed.: DSM-IV*. Washington, DC: American Psychiatric Association.

APA (2000). *Diagnostic and Statistical Manual of Mental Disorders, 4th ed. text revision: DSM-IV-TR*. Washington, DC: American Psychiatric Association.

APA (2013). *Diagnostic and Statistical Manual of Mental Disorders, 5th ed.: DSM-5*. Washington, DC: American Psychiatric Association.

Aragona, Massimiliano (2009). The role of comorbidity in the crisis of the current psychiatric classification system. *Philosophy, Psychiatry, & Psychology*, 16(1), 1–11.

Bateson, Gregory, Jackson, Don D., Haley, Jay, & Weakland, John (1956). Towards a theory of schizophrenia. *Behavioral Science*, 1(4), 251–4.

Bechtel, William, & Abrahamsen, Adele (2005). Explanation: A mechanistic alternative. *Studies in History and Philosophy of Biological and Biomedical Sciences*, 36(2), 421–41.

Bechtel, William, & Richardson, Robert C. (2010). *Discovering Complexity: Decomposition and Localization as Strategies in Scientific Research*, 2nd ed. Cambridge, MA: MIT Press.

Beebee, Helen, & Sabbarton-Leary, Nigel. (2010). Are psychiatric kinds "real"? *European Journal of Analytic Philosophy*, 6(1), 11–27.

Bird, Alexander, & Tobin, Emma (2018). Natural kinds. In Edward N. Zalta, ed., *Stanford Encyclopedia of Philosophy* (Spring 2018 ed.). https://plato.stanford.edu/archives/spr2018/entries/natural-kinds/

Blashfield, Roger K. (1984). *The Classification of Psychopathology: Neo-Kraepelinian and Quantitative Approaches*. New York: Plenum Press.

Blease, Charlotte (2010). Scientific progress and the prospects for culture-bound syndromes. *Studies in History and Philosophy of Biological and Biomedical Sciences*, 41(4), 333–9.

Bluhm, Robyn (2017). The need for new ontologies in psychiatry. *Philosophical Explorations*, 20(2), 146–59.

Bogen, James (1988). Comments on "The sociology of knowledge about child abuse." *Noûs*, 22(1), 65–6.

Boorse, Christopher (1975). On the distinction between disease and illness. *Philosophy and Public Affairs*, 5(1), 49–68.

Boorse, Christopher (1976a). What a theory of mental health should be. *Journal for the Theory of Social Behaviour*, 6(1), 61–84.

Boorse, Christopher (1976b). Wright on functions. *Philosophical Review*, 85(1), 70–86.

Boorse, Christopher (1977). Health as a theoretical concept. *Philosophy of Science*, 44(4), 542–73.

Boorse, Christopher (1997). A rebuttal on health. In James M. Humber & Robert F. Almeder, eds., *What Is Disease?* Totowa, NJ: Humana Press, pp. 3–143.

Boorse, Christopher (2014). A second rebuttal on health. *Journal of Medicine and Philosophy*, 39(6), 683–724.

Borsboom, Denny, & Cramer, Angélique O. J. (2013). Network analysis: An integrative approach to the structure of psychopathology. *Annual Review of Clinical Psychology*, 9, 91–121.

Boyd, Richard (1991). Realism, anti-foundationalism and the enthusiasm for natural kinds. *Philosophical Studies*, 61(1–2), 127–48.

Boyd, Richard (1999a). Homeostasis, species, and higher taxa. In Robert A. Wilson, ed., *Species: New Interdisciplinary Essays*. Cambridge, MA: MIT Press, pp. 141–86.

Boyd, Richard (1999b). Kinds, complexity, and multiple realization: Comments on Millikan's "Historical kinds and the special sciences." *Philosophical Studies*, 95(1–2), 67–98.

Bracken, Patrick, & Thomas, Philip (2005). *Postpsychiatry: Mental Health in a Postmodern World*. Oxford: Oxford University Press.

Bracken, Patrick, Thomas, Philip, and Timimi, Sami, et al. (2012). Psychiatry beyond the current paradigm. *British Journal of Psychiatry*, 201(6), 430–4.

Bueter, Anke (2019a). A multi-dimensional pluralist response to the DSM-controversies. *Perspectives on Science*, 27(2), 316–43.

Bueter, Anke (2019b). Epistemic injustice and psychiatric classification. *Philosophy of Science*, 86(5), 1064–74.

Burmeister, Margit, McInnis, Melvin G., & Zöllner, Sebastian (2008). Psychiatric genetics: Progress amid controversy. *Nature Reviews Genetics*, 9(7), 527–40.

Buss, David M., Haselton, Martie, Shackelford, Todd K., Bleske, April L., & Wakefield, Jerome C. (1988). Adaptations, exaptations, and spandrels. *American Psychologist*, 53(5), 553–48.

Carnap, Rudolf (1950). *Logical Foundations of Probability*. Chicago: University of Chicago Press.

Cartwright, Nancy (1983). *How the Laws of Physics Lie*. Oxford: Oxford University Press.

Chalmers, David J. (1996). *The Conscious Mind: In Search of a Fundamental Theory*. Oxford: Oxford University Press.

Cooper, Rachel (2005). *Classifying Madness: A Philosophical Examination of the Diagnostic and Statistical Manual of Mental Disorders*. Dordrecht: Springer.

Cooper, Rachel (2007). *Psychiatry and Philosophy of Science*. Chesham: Acumen.

Cooper, Rachel (2010). Are culture-bound syndromes as real as universally-occurring disorders? *Studies in History and Philosophy of Biological and Biomedical Sciences*, 41(4), 325–32.

Cooper, Rachel (2013). Natural kinds. In K. W. M. Fulford, Martin Davies, Richard G., and T. Gipps, et al., eds., *The Oxford Handbook of Philosophy and Psychiatry*. Oxford: Oxford University Press, pp. 950–65.

Craver, Carl F. (2009). Mechanisms and natural kinds. *Philosophical Psychology*, 22(5), 575–94.

Craver, Carl, & Tabery, James (2019). Mechanisms in science. In Edward N. Zalta, ed., *Stanford Encyclopedia of Philosophy* (Summer 2019 ed.). https://plato.stanford.edu/archives/sum2019/entries/science-mechanisms/

Cronbach, Lee J., & Meehl, Paul E. (1955). Construct validity in psychological tests. *Psychological Bulletin*, 52(4), 281–302.

Crow, T. J. (1980). Molecular pathology of schizophrenia: More than one disease process? *British Medical Journal*, 280(6207), 1–9.

Culp, Sylvia (1995). Objectivity in experimental inquiry: Breaking data-technique circles. *Philosophy of Science*, 62(3), 430–50.

Cummins, Robert (1975). Functional analysis. *Journal of Philosophy*, 72(20), 741–65.

Cuthbert, Bruce, & Insel, Thomas (2010). The data of diagnosis: New approaches to psychiatric classification. *Psychiatry*, 73(4), 311–14.

Davidson, Donald (1970). Mental events. In Lawrence Foster & J. W. Swanson, eds., *Experience and Theory*. Amherst: University of Massachusetts Press, pp. 79–101.

Davidson, Richard J., Pizzagalli, Diego, Nitschke, Jack B., & Putnam, Katherine (2002). Depression: Perspectives from affective neuroscience. *Annual Review of Psychology*, 53, 545–74.

DePaul, Michael, & Ramsey, William (eds.) (1998). *Rethinking Intuition: The Psychology of Intuition and Its Role in Philosophical Inquiry*. Lanham, MD: Rowman & Littlefield.

Devitt, Michael (2008). Resurrecting biological essentialism. *Philosophy of Science*, 75(3), 344–82.

Devitt, Michael (2010). Species have (partly) intrinsic essences. *Philosophy of Science*, 77(5), 648–61.

Douglas, Heather E. (2009). *Science, Policy, and the Value-Free Ideal*. Pittsburgh, PA: University of Pittsburgh Press.

Drevets, Wayne C., Thase, Michael E., Moses-Kolko, Eydie L., et al. (2007). Serotonin-1A receptor imaging in recurrent depression: Replication and literature review. *Nuclear Medicine and Biology*, 34(7), 865–77.

Dupré, John (1993). *The Disorder of Things: Metaphysical Foundations of the Disunity of Science*. Cambridge, MA: Harvard University Press.

Dupré, John (2001). In defence of classification. *Studies in History and Philosophy of Biological and Biomedical Sciences*, 32(2), 203–19.

Engelhardt, H. Tristam (1976). Ideology and etiology. *Journal of Medicine and Philosophy*, 1(3), 256–68.

Ereshefsky, Marc (2009). Defining "health" and "disease." *Studies in History and Philosophy of Biological and Biomedical Sciences*, 40(3), 221–7.

Ereshefsky, Marc (2010). What's wrong with the new biological essentialism? *Philosophy of Science*, 77(5), 674–85.

Ereshefsky, Marc (2017). Species. In Edward N. Zalta, ed., *Stanford Encyclopedia of Philosophy* (Fall 2017 ed.). https://plato.stanford.edu/arch ives/fall2017/entries/species/

Faucher, Luc, & Goyer, Simon (2015). RDoC: Thinking outside the DSM box without falling into a reductionist trap. In Steeves Demazeux & Patrick Singy, eds., *The DSM-5 in Perspective: Philosophical Reflections on the Psychiatric Babel*. Dordrecht: Springer, pp. 199–224.

Feighner, John P., Robins, Eli, Guze, Samuel B., et al. (1972). Diagnostic criteria for use in psychiatric research. *Archives of General Psychiatry*, 26(1), 57–63.

First, Michael B., Pincus, Harold A., Levine, John B., et al. (2004). Clinical utility as a criterion for revising psychiatric diagnoses. *American Journal of Psychiatry*, 161(6), 946–54.

First, Michael B., & Westen, Drew (2007). Classification for clinical practice: How to make ICD and DSM better able to serve clinicians. *International Review of Psychiatry*, 19(5), 473–81.

Frances, Allen J., & Widiger, Thomas (2012). Psychiatric diagnosis: Lessons from the DSM-IV past and cautions for the DSM future. *Annual Review of Clinical Psychology*, 8, 109–30.

Franklin, Allan (1996). There are no antirealists in the laboratory. In Robert S. Cohen, Risto Hilpinen, & Qiu Renzong, eds., *Realism and Antirealism in the Philosophy of Science*. Dordrecht: Kluwer Academic, pp. 131–48.

Franklin, Allan, & Howson, Colin (1984). Why do scientists prefer to vary their experiments? *Studies in History and Philosophy of Science*, 15(1), 51–62.

Franklin, Allan, & Perovic, Slobodan (2019). Experiment in physics. In Edward N. Zalta, ed., *Stanford Encyclopedia of Philosophy* (Winter 2019 ed.). https://plato.stanford.edu/archives/win2019/entries/physics-experiment/

Fulford, K. W. M., Davies, Martin, and Gipps, Richard G. T., et al. (eds.) (2013). *The Oxford Handbook of Philosophy and Psychiatry*. Oxford: Oxford University Press.

Garson, Justin (2019). *What Biological Functions Are and Why They Matter*. Cambridge: Cambridge University Press.

Glennan, Stuart (2002). Rethinking mechanistic explanation. *Philosophy of Science*, 69(S3), S342–S353.

Glennan, Stuart (2017). *The New Mechanical Philosophy*. Oxford: Oxford University Press.

Goffman, Erving (1961). *Asylums: Essays on the Social Situation of Mental Patients and Other Inmates*. Garden City, NY: Anchor.

Goldstein, Jan E. (1989). *To Console and Classify: The French Psychiatric Profession in the Nineteenth Century*. Chicago: University of Chicago Press.

Goodman, Nelson (1955). *Fact, Fiction, and Forecast*. Cambridge, MA: Harvard University Press.

Gould, Stephen J. (1991). Exaptation: A crucial tool for evolutionary analysis. *Journal of Social Issues*, 47(3), 43–65.

Gould, Stephen J., & Lewontin, Richard (1979). The spandrels of San Marco and the Panglossian paradigm: A critique of the adaptationist programme. *Proceedings of the Royal Society, London*, B205(1161), 581–98.

Gould, Stephen J., & Vrba, Elisabeth S. (1982). Exaptation: A missing term in the science of form. *Paleobiology*, 8(1), 4–15.

Griffiths, Paul E. (1999). Squaring the circle: Natural kinds with historical essences. In Robert A. Wilson, ed., *Species: New Interdisciplinary Essays*. Cambridge, MA: MIT Press, pp. 208–28.

Guillin, Olivier, Abi-Dhargham, Anissa, & Laruelle, Marc (2007). Neurobiology of dopamine in schizophrenia. In Anissa Abi-Dhargham & Olivier Guillin, eds., *Integrating the Neurobiology of Schizophrenia*. San Diego, CA: Elsevier, pp. 1–39.

Gupta, Anil (2019). Definitions. In Edward N. Zalta, ed., *Stanford Encyclopedia of Philosophy* (Winter 2019 ed.). https://plato.stanford.edu/archives/win2019/entries/definitions/

Guze, Samuel B. (1992). *Why Psychiatry Is a Branch of Medicine*. New York: Oxford University Press.

Hacking, Ian (1983). *Representing and Intervening: Introductory Topics in the Philosophy of Natural Science*. Cambridge: Cambridge University Press.

Hacking, Ian (1995a). *Rewriting the Soul: Multiple Personality and the Sciences of Memory*. Princeton, NJ: Princeton University Press.

Hacking Ian (1995b). The looping effects of human kinds. In Dan Sperber, David Premack, & Anne J. Premack, eds., *Causal Cognition: A Multidisciplinary Debate*. Oxford: Clarendon Press, pp. 351–83.

Hacking, Ian (1998). *Mad Travelers: Reflections on the Reality of Transient Mental Illness*. Charlottesville: University Press of Virginia.

Hacking, Ian (1999). *The Social Construction of What?* Cambridge, MA: Harvard University Press.

Hacking, Ian (2007). Kinds of people: Moving targets. *Proceedings of the British Academy*, 151, 285–318.

Hacking, Ian (2013). Lost in the forest. Review of *DSM-5: Diagnostic and Statistical Manual of Mental Disorders, Fifth Edition* by the American Psychiatric Association. *London Review of Books*, 35(15), 7–8.

Hancock, Stephanie D., & McKim, William A. (2018). *Drugs and Behavior: An Introduction to Behavioral Pharmacology*, 8th ed. New York: Pearson.

Haslam, Nick (2003). Kinds of kinds: A conceptual taxonomy of psychiatric categories. *Philosophy, Psychiatry, & Psychology*, 9(3), 203–17.

Haslanger, Sally (2012). *Resisting Reality: Social Construction and Social Critique*. Oxford: Oxford University Press.

Hausman, Daniel M. (2012). Health, naturalism, and functional efficiency. *Philosophy of Science*, 79(4), 519–41.

Heim, Christine, Newport, D. Jeffrey, Mleztko, Tanja, Miler, Andrew H., & Nemereroff, Charles B. (2008). The link between childhood trauma and depression: Insights from the HPA axis studies in humans. *Psychoneuroendocrinology*, 33(6), 693–710.

Hempel, Carl G. (1965). *Aspects of Scientific Explanation and Other Essays in the Philosophy of Science*. New York: Free Press.

Hirschfeld, Robert M. A. (1994). Major depression, dysthymia, and depressive personality disorder. *British Journal of Psychiatry*, 165(S26), 23–30.

Hoffman, Ginger A., & Zachar, Peter (2017). RDoC's metaphysical assumptions: Problems and promises. In Jeffrey Poland & Şerife Tekin, eds., *Extraordinary Science and Psychiatry: Response to the Crisis in Mental Health Research*. Cambridge, MA: MIT Press, pp. 59–86.

Horwitz, Allan V., & Wakefield, Jerome C. (2007). *The Loss of Sadness: How Psychiatry Transformed Normal Sorrow into Depressive Disorder*. Oxford: Oxford University Press.

Howes, Oliver D., & Kapur, Shitij (2009). The dopamine hypothesis of schizophrenia: Version III: The final common pathway. *Schizophrenia Bulletin*, 35(3), 549–62.

Hyman, Steven E. (2002). Neuroscience, genetics, and the future of psychiatric diagnosis. *Psychopathology*, 35(2–3), 139–44.

Hyman, Steven E. (2007). Can neuroscience be integrated into the DSM-V? *Nature Reviews Neuroscience*, 8(9), 725–32.

Hyman, Steven E. (2010). The diagnosis of mental disorders: The problem of reification. *Annual Review of Clinical Psychology*, 6, 155–79.

Hyman, Steven E., & Fenton, Wayne S. (2003). What are the right targets for psychopharmacology? *Science*, 299(5605), 350–1.

Insel, Thomas, Cuthbert, Bruce, Garvey, Marjorie, et al. (2010). Research Domain Criteria (RDoC): Toward a new classification framework for research on mental disorders. *American Journal of Psychiatry*, 167(7), 748–51.

Insel, Thomas R., & Lieberman, Jeffrey A. (2013). DSM-5 and RDoC: Shared interests. Joint press release by the National Institute of Mental Health and the American Psychiatric Association (March 13, 2013). http://www.nimh.nih.gov/archive/news/2013/dsm-5-and-rdoc-shared-interests.shtml

Jablensky, Assen (2016). Psychiatric classifications: Validity and utility. *World Psychiatry*, 15(1), 26–31.

Jackson, Frank (1998). *From Ethics to Metaphysics*. Oxford: Oxford University Press.

Javitt, Daniel C. (2007). Glutamate and schizophrenia: Phencyclidine, N-methyl-$_D$-aspartate receptors, and dopamine-glutamate interactions. In Anissa Abi-Dhargham & Olivier Guillin, eds., *Integrating the Neurobiology of Schizophrenia*. San Diego, CA: Elsevier, pp. 69–108.

Kendell, Robert E. (1975). The concept of disease and its implications for psychiatry. *British Journal of Psychiatry*, 127(4), 305–15.

Kendell, Robert, & Jablensky, Assen (2003). Distinguishing between the validity and utility of psychiatric diagnoses. *American Journal of Psychiatry*, 160(1), 4–12.

Kendler Kenneth S. (1980). The nosologic validity of paranoia (simple delusional disorder): A review. *Archives of General Psychiatry*, 37(6), 699–706.

Kendler, Kennth S., & First, Michael B. (2010). Alternative futures for the DSM revision process: Iteration *v*. paradigm shift. *British Journal of Psychiatry*, 197(4), 263–5.

Kendler, Kenneth S., Prescott, Carol A., Myers, John, & Neale, Michael C. (2003). The structure of genetic and environmental risk for common psychiatric and substance use disorders in men and women. *Archives of General Psychiatry*, 60(9), 929–37.

Kendler, Kenneth S., & Schaffner, Kenneth F. (2011). The dopamine hypothesis of schizophrenia: An historical and philosophical analysis. *Philosophy, Psychiatry, & Psychology*, 18(1), 41–63.

Kendler, Kenneth S., Zachar, Peter, & Craver, Carl (2011). What kinds of things are psychiatric disorders? *Psychological Medicine*, 41(6), 1143–50.

Kessler, Ronald C., Berglund, Patricia, Demler, Olga, et al. (2003). The epidemiology of depression: Results from the National Comorbidity Survey Replication (NCS-R). *Journal of the American Medical Association*, 289(23), 3095–105.

Kessler, Ronald C., Berglund, Patricia, Demler, Olga, et al. (2005). Lifetime prevalence and age-of-onset distributions of *DSM-IV* disorders in the National Comorbidity Survey Replication. *Archives of General Psychiatry*, 62(6), 593–602.

Khalidi, Muhammad Ali (1998). Natural kinds and cross-cutting categories. *Journal of Philosophy*, 95(1), 33–50.

Khalidi, Muhammad Ali (2010). Interactive kinds. *British Journal for the Philosophy of Science*, 61(1), 335–60.

Khalidi, Muhammad Ali (2013). *Natural Categories and Human Kinds: Classification in the Natural and Social Sciences*. Cambridge: Cambridge University Press.

Kieseppä, Tuula, Partonen, Timo, Haukka, Jari, Kaprio, Jakko, & Lönnqvist, Jouko (2004). High concordance of bipolar I disorder in a nationwide sample of twins. *American Journal of Psychiatry*, 161(10), 1814–21.

Kim, Yunjung, Zerwas, Stephanie, Trace, Sara E., & Sullivan, Patrick F. (2011). Schizophrenia genetics: Where next? *Schizophrenia Bulletin*, 37(3), 456–63.

Kincaid, Harold, Dupré, John, & Wylie, Alison (eds.) (2007). *Value Free Science? Ideals and Illusions*. Oxford: Oxford University Press.

Kingma, Elselijn (2007). What is it to be healthy? *Analysis*, 67(2), 128–33.

Kingma, Elselijn (2010). Paracetamol, poison, and polio: Why Boorse's account of function fails to distinguish health and disease. *British Journal for the Philosophy of Science*, 61(2), 241–64.

Kious, Brent M. (2018). Boorse's theory of disease: (Why) do values matter? *Journal of Medicine and Philosophy*, 43(4), 421–38.

Kirmayer, Laurence J. (2001). Cultural variations in the clinical presentation in depression and anxiety: Implications for diagnosis and treatment. *Journal of Clinical Psychiatry*, 13(Suppl. 13), 22–8.

Kleinman, Arthur (1988). *Rethinking Psychiatry: From Cultural Category to Personal Experience*. New York: Free Press.

Kotov, Roman, Krueger, Robert F., Watson, David, et al. (2017). The hierarchical taxonomy of psychopathology (HiTOP): A dimensional alternative to traditional nosologies. *Journal of Abnormal Psychology*, 126(4), 454–77.

Kring, Ann M., Davison, Gerald, Neal, John M., & Johnson, Sheri L. (2017). *Abnormal Psychology: The Science and Treatment of Psychological Disorders*, 13th ed. Hoboken, NJ: John Wiley & Sons.

Kupfer, David J., First, Michael B., & Regier, Darrel A. (2002). Introduction. In David J. Kupfer, Michael B. First, & Darrel A. Regier, eds., *A Research Agenda for DSM-V*. Washington, DC: American Psychiatric Association, pp. xv–xxiii.

Kupfer, David J., & Regier, Darrel A. (2011). Neuroscience, clinical evidence, and the future of psychiatric classification in *DSM-5*. *American Journal of Psychiatry*, 168(7), 672–4.

Kutschenko, Lara K. (2011). How to make sense of broadly applied medical classification systems: Introducing epistemic hubs. *History and Philosophy of the Life Sciences*, 33(4), 583–602.

Laing, R. D. (1967). *The Politics of Experience*. New York: Ballatine.

Laruelle, Marc, Kegeles, Lawrence S., & Abi-Dargham, Anissa (2003). Glutamate, dopamine, and schizophrenia: From pathology to treatment. *Annals of the New York Academy of Sciences*, 1003(1), 138–58.

Laurence, Stephen, & Margolis, Eric (2003). Concepts and conceptual analysis. *Philosophy and Phenomenological Research*, 67(2), 253–82.

Lemoine, Maël (2013). Defining disease beyond conceptual analysis: An analysis of conceptual analysis in philosophy of medicine. *Theoretical Medicine and Bioethics*, 34(4), 309–25.

Lewis, David A. (2012). Cortical circuit dysfunction and cognitive deficits in schizophrenia – implications for preemptive interventions. *European Journal of Neuroscience*, 35(12), 1871–8.

Lewis, David A., & González-Burgos, Guillermo (2008). Neuroplasticity of neocortical circuits in schizophrenia. *Neuropsychopharmacology Reviews*, 33(1), 141–65.

Lewontin, Richard C. (1979). Sociobiology as an adaptationist program. *Behavioral Sciences*, 24(1), 5–14.

Lilienfeld, Scott O., & Marino, Lori (1995). Mental disorder as a Roschian concept: A critique of Wakefield's "harmful dysfunction" analysis. *Journal of Abnormal Psychology*, 104(3), 411–20.

Lilienfeld, Scott O., & Marino, Lori (1999). Essentialism revisited: Evolutionary theory and mental disorders. *Journal of Abnormal Psychology*, 108(3), 400–11.

Lin, Keh-Ming, Poland, Russell E., & Anderson, Dora (1995). Psychopharmacology, ethnicity and culture. *Transcultural Psychiatric Research Review*, 32(1), 3–40.

Lloyd, Elisabeth A. (1999). Evolutionary psychology: The burdens of proof. *Biology & Philosophy*, 14(2), 211–33.

Longino, Helen E. (1990). *Science As Social Knowledge: Values and Objectivity in Scientific Inquiry*. Princeton, NJ: Princeton University Press.

Machamer, Peter, Darden, Lindley, & Craver, Carl F. (2000). Thinking about mechanisms. *Philosophy of Science*, 67(1), 1–25.

Machery, Edouard (2017). *Philosophy within Its Proper Bounds*. Oxford: Oxford University Press.

Maj, Mario (2005). "Psychiatric co-morbidity": An artefact of current diagnostic systems. *British Journal of Psychiatry*, 186(3), 182–4.

Major Depressive Disorder Working Group of the Psychiatric GWAS Consortium (2013). A mega-analysis of genome-wide association studies for major depressive disorder. *Molecular Psychiatry*, 18(4), 497–511.

Mallon, Ron (2016). *The Construction of Human Kinds*. Oxford: Oxford University Press.

Marsella, Anthony J. (1988). Cross-cultural research on severe mental disorders: Issues and findings. *Acta Psychiatrica Scandinavica*, 78(S344), 7–22.

Maung, Hane Htut (2016). Diagnosis and causal explanation in psychiatry. *Studies in History and Philosophy of Biological and Biomedical Sciences*, 60, 15–29.

Mayes, Rick, & Horwitz, Allan V. (2005). DSM-III and the revolution in the classification of mental illness. *Journal of the History of the Behavioral Sciences*, 41(3), 249–67.

McClung, Colleen A. (2013). How might circadian rhythms control mood? Let me count the ways. *Biological Psychiatry*, 74(4), 242–9.

McGrath, John, Saha, Sakanta, Chant, David, & Welham, Joy (2008). Schizophrenia: A concise overview of incidence, prevalence, and mortality. *Epidemiologic Reviews*, 30(1), 67–76.

Mead, George H. (1934). *Mind, Self, and Society: From the Standpoint of a Social Behaviorist*. Charles W. Morris, ed. Chicago: University of Chicago Press.

Meehl, Paul E. (1973). *Psychodiagnosis: Selected Papers*. Minneapolis: University of Minnesota Press.

Meehl, Paul E. (1991). *Selected Philosophical and Methodological Papers*. C. Anthony Anderson & Keith Gunderson, eds. Minneapolis: University of Minnesota Press.

Merikangas, Kathleen R., Akiskal, Hagop S., Angst, Jules, et al. (2007). Lifetime and 12-month prevalence of bipolar spectrum disorder in the National Comorbidity Survey Replication. *Archives of General Psychiatry*, 64(5), 543–52.

Messick, Samuel (1995). Validity of psychological assessment: Validation of inferences from persons' responses and performances as scientific inquiry into score meaning. *American Psychologist*, 50(9), 741–9.

Micale, Mark S. (1993). On the "disappearance" of hysteria: A study in the clinical deconstruction of a disorder. *Isis*, 84(3), 496–526.

Middleton, Hugh, & Moncrieff, Jonanna (2019). Critical psychiatry: A brief overview. *BJPsych Advances*, 25(1), 47–54.

Millikan, Ruth G. (1989). In defense of proper functions. *Philosophy of Science*, 56(2), 288–302.

Millikan, Ruth G. (1999). Historical kinds and the "special sciences." *Philosophical Studies*, 95(1–2), 45–65.

Mitchell, Sandra D. (2003). *Biological Complexity and Integrative Pluralism*. Cambridge: Cambridge University Press.

Mitchell, Sandra D. (2008). Comment: Taming causal complexity. In Kenneth S. Kendler & Josef Parnas, eds., *Philosophical Issues in Psychiatry: Explanation, Phenomenology, and Nosology*. Baltimore, MD: Johns Hopkins University Press, pp. 125–31.

Moffitt, Terrie E., Caspi, Avshalom, Harrington, Honallee, et al. (2007). Generalized anxiety disorder and depression: Childhood risk factors in a birth cohort followed to age 32. *Psychological Medicine*, 37(3), 441–52.

Moghaddam, Bita, & Javitt, Daniel (2012). From revolution to evolution: The glutamate hypothesis of schizophrenia and its implications for treatment. *Neuropsychopharmacology Reviews*, 37(1), 4–15.

Moret, Chantal, & Briley, Michael (2011). The importance of norepinephrine in depression. *Neuropsychiatric Disease and Treatment*, 7(Supp.1), 9–13.

Morris, Sarah E., & Cuthbert, Bruce N. (2012). Research Domain Criteria: Cognitive systems, neural networks, and dimensions of behavior. *Dialogues in Clinical Neuroscience*, 14(1), 29–37.

Murphy, Dominic (2006). *Psychiatry in the Scientific Image*. Cambridge, MA: MIT Press.

Murphy, Dominic (2020a). Concepts of disease and health. In Edward N. Zalta, ed., *Stanford Encyclopedia of Philosophy* (Summer 2020 ed.). https://plato.stanford.edu/archives/sum2020/entries/health-disease/

Murphy, Dominic (2020b). Philosophy of psychiatry. In Edward N. Zalta, ed., *Stanford Encyclopedia of Philosophy* (Fall 2020 ed.). https://plato.stanford.edu/archives/fall2020/entries/psychiatry/

Murphy, Dominic, & Stich, Stephen (2000). Darwin in the madhouse: Evolutionary psychology and the classification of mental disorder. In Peter Carruthers & Andrew Chamberlain, eds., *Evolution and the Human Mind: Modularity, Language and Meta-Cognition*. Cambridge: Cambridge University Press, pp. 62–92.

Murphy, Dominic, & Woolfolk, Robert L. (2000). The harmful dysfunction analysis of mental disorder. *Philosophy, Psychiatry, & Psychology*, 7(4), 241–52.

Neander, Karen (1991a). Functions as selected effects: The conceptual analysts' defense. *Philosophy of Science*, 58(2), 168–84.

Neander, Karen (1991b). The teleological notion of "function." *Australasian Journal of Philosophy*, 69(4), 454–68.

Nelson-Gray, Rosemery O. (1991). *DSM-IV*: Empirical guidelines from psychometrics. *Journal of Abnormal Psychology*, 100(3), 308–15.

Nestler, Eric J., & Carlezon Jr., William A. (2006). The mesolimbic dopamine reward circuit in depression. *Biological Psychiatry*, 59(12), 1151–9.

NIMH (2018). RDoC Matrix Archives. National Institute of Mental Health. www.nimh.nih.gov/research/research-funded-by-nimh/rdoc/constructs/rdoc-matrix-archives.shtml

Olney, John W., & Farber, Nuri B. (1995). Glutamate receptor dysfunction and schizophrenia. *Archives of General Psychiatry*, 52(12), 998–1007.

Padovani, Flavia, Richardson, Alan, & Tsou, Jonathan Y. (eds.) (2015). *Objectivity in Science: New Perspectives from Science and Technology Studies*. Cham: Springer.

Parnas, Josef (2015). Differential diagnosis and current polythetic classification. *World Psychiatry*, 14(3), 284–7.

Perreault, Charles (2012). The pace of cultural evolution. *PLoS ONE*, 7(9), e45150. www.ncbi.nlm.nih.gov/pmc/articles/PMC3443207/

Pittenger, Christopher, & Duman, Ronald S. (2008). Stress, depression, and neuroplasticity: A convergence of mechanisms. *Neuropsychopharmacology Reviews*, 33(1), 88–109.

Powell, Russell, & Scarffe, Eric (2019). Rethinking "disease": A fresh new diagnosis and a new philosophical treatment. *Journal of Medical Ethics*, 45(9), 579–88.

Putnam, Hilary (1960). Minds and machines. In Sidney Hook, ed., *Dimensions in Mind*. New York: New York University Press, pp. 138–64.

Quine, W. V. O. (1960). *Word and Object*. Cambridge, MA: MIT Press.

Quine, W. V. O. (1969). *Ontological Relativity and Other Essays*. New York: Columbia University Press.

Radden, Jennifer (2009). *Moody Minds Distempered: Essays on Melancholy and Depression*. Oxford: Oxford University Press.

Radden, Jennifer (2019). Mental disorder (illness). In Edward N. Zalta ed., *Stanford Encyclopedia of Philosophy* (Winter 2019 ed.). https://plato.stanford.edu/archives/win2019/entries/mental-disorder/

Ramsey, William (1992). Prototypes and conceptual analysis. *Topoi*, 11(1), 59–70.

Regier, Darrel A., Narrow, William E., Kuhl, Emily A., & Kupfer, David J. (2009). The conceptual development of DSM-V. *American Journal of Psychiatry*, 166(6), 645–50.

Reiss, Julian & Ankeny, Rachel A. (2016). Philosophy of medicine. In Edward N. Zalta, ed., *Stanford Encyclopedia of Philosophy* (Summer 2016 ed.). https://plato.stanford.edu/archives/sum2016/entries/medicine/

Reiss, Julian, & Jan Sprenger (2020). Scientific objectivity. In Edward N. Zalta, ed., *Stanford Encyclopedia of Philosophy* (Winter 2020 ed.). https://plato.stanford.edu/archives/win2020/entries/scientific-objectivity/

Rhebergen, Didi, & Graham, Rebecca (2014). The re-labelling of dysthymic disorder to persistent depressive disorder in DSM-5: Old wine in new bottles? *Current Opinion in Psychiatry*, 27(1), 27–31.

Richerson, Peter J., Boyd, Robert, & Henrich, Joseph (2010). Gene-culture coevolution in the age of genomics. *Proceedings of the National Academy of Sciences of the United States of America*, 107(Suppl. 2), 8985–92.

Robins, Eli, & Guze, Samuel B. (1970). Establishment of diagnostic validity in psychiatric illness: Its application to schizophrenia. *American Journal of Psychiatry*, 126(7), 983–6.

Ross, Lauren N. (2021). Causal concepts in biology: How pathways differ from mechanisms and why it matters. *British Journal for the Philosophy of Science*, 72(1), 131–58.

Sadler, John Z. (2013). Considering the economy of DSM alternatives. In Joel Paris & James Phillips, eds., *Making the DSM-5: Concepts and Controversies*. New York: Springer, pp. 21–38.

Saha, Sukanta, Chant, David, Welham, Joy, & McGrath, John (2005). A systematic review of the prevalence of schizophrenia. *PLoS Medicine*, 2 (5), e141. https://doi.org/10.1371/journal.pmed.0020141

Sanislow, Charles A., Pine, Daniel S., Quinn, Kevin J., et al. (2010). Developing constructs for psychopathology research: Research Domain Criteria. *Journal of Abnormal Psychology*, 119(4), 631–9.

Sartorius, N., Davidian, H., Ernberg, G., et al. (1983). *Depressive Disorders in Different Cultures: Report on the WHO Collaborative Study on Standardized Assessment of Depressive Disorders*. Geneva: World Health Organization.

Scadding, J. G. (1990). The semantic problems of psychiatry. *Psychological Medicine*, 20(2), 243–8.

Schaffner, Kenneth F. (1993). *Discovery and Explanation in Biology and Medicine*. Chicago: University of Chicago Press.

Schaffner, Kenneth F. (2012). A philosophical overview of the problems of validity for psychiatric disorders. In Kenneth S. Kendler & Josef Parnas, eds., *Philosophical Issues in Psychiatry II: Nosology*. Oxford: Oxford University Press, pp. 169–89.

Schaffner, Kenneth F. (2016). *Behaving: What's Genetic, What's Not, and Why Should We Care?* Oxford: Oxford University Press.

Scheff, Thomas J. (1963). The role of the mentally ill and the dynamics of disorder: A research framework. *Sociometry*, 26(4), 436–53.

The Schizophrenia Psychiatric Genome-Wide Association Study (GWAS) Consortium (2011). Genome-wide association study identifies five new schizophrenia loci. *Nature Genetics*, 43(10), 969–76.

Schuham, Anthony I. (1967). The double bind hypothesis a decade later. *Psychological Bulletin*, 68(6), 409–16.

Schupbach, Jonah N. (2018). Robustness analysis as explanatory reasoning. *British Journal for the Philosophy of Science*, 69(1), 275–300.

Schwartz, Michael A., & Wiggins, Osborne P. (1987). Diagnosis and ideal types: A contribution to psychiatric classification. *Comprehensive Psychiatry*, 28(4), 277–91.

Schwartz, Peter H. (2007a). Decision and discovery in defining "disease." In Harold Kincaid & Jennifer McKitrick, eds., *Establishing Medical Reality*. Dordrecht: Springer, pp. 47–63.

Schwartz, Peter H. (2007b). Defining dysfunction: Natural selection, design, and drawing a line. *Philosophy of Science*, 74(3), 364–85.

Schwartz, Peter H. (2014). Reframing the disease debate and defending the biostatistical theory. *Journal of Medicine and Philosophy*, 39(6), 572–89.

Sedgwick, Peter (1973). Illness: Mental and otherwise. *Hastings Center Studies*, 1(3), 19–40.

Soler, Léna, Trizio, Emiliano, Nickles, Thomas, & Wimsatt, William C. (eds.) (2012). *Characterizing the Robustness of Science: After the Practice Turn in Philosophy of Science*. Dordrecht: Springer.

Solomon, Miriam (2015). Expert disagreement and medical authority. In Kenneth S. Kendler & Josef Parnas, eds., *Philosophical Issues in Psychiatry III: The Nature and Sources of Historical Change*. Oxford: Oxford University Press, pp. 60–72.

Stahl, Stephen M. (2002). The psychopharmacology of energy and fatigue. *Journal of Clinical Psychiatry*, 63(1), 7–8.

Stegenga, Jacob (2009). Robustness, discordance, and relevance. *Philosophy of Science*, 76(5), 650–61.

Stegenga, Jacob (2018). *Medical Nihilism*. Oxford: Oxford University Press.

Stich, Stephen (1992). What is a theory of mental representation? *Mind*, 101(402), 243–61.

Sullivan, Jaqueline A. (2014). Stabilizing mental disorders: Prospects and problems. In Harold Kincaid & Jaqueline A. Sullivan, eds., *Classifying Psychopathology: Mental Kinds and Natural Kinds*. Cambridge, MA: MIT Press, pp. 257–82.

Sullivan, Jaqueline A. (2017). Coordinated pluralism as a means to facilitate integrative taxonomies of cognition. *Philosophical Explorations*, 20(2), 129–45.

Sullivan, Patrick F. (2005). The genetics of schizophrenia. *PLoS Medicine*, 2(7), e212. https://doi.org/10.1371/journal.pmed.0020212

Sullivan, Patrick F., Daly, Mark J., & O'Donovan, Michael (2012). Genetic architectures of psychiatric disorders: The emerging picture and its implications. *Nature Reviews Genetics*, 13(8), 537–51.

Sullivan, Patrick F., & Kendler, Kenneth S. (1998). Typology of common psychiatric syndromes: An empirical study. *British Journal of Psychiatry*, 173(4), 312–19.

Sullivan, Patrick F., Neale, Michael C., & Kendler, Kenneth S. (2000). Genetic epidemiology of major depression: Review and meta-analysis. *American Journal of Psychiatry*, 157(10), 1552–62.

Szasz, Thomas S. (1960). The myth of mental illness. *American Psychologist*, 15(2), 113–18.

Szasz, Thomas S. (1974). *The Myth of Mental Illness: Foundations of a Theory of Personal Conduct*, 2nd ed. New York: Harper & Row.

Szasz, Thomas (1988). *Schizophrenia: Psychiatry's Sacred Symbol*. Syracuse: Syracuse University Press.

Szasz, Thomas (1997). *Insanity: The Idea and Its Consequences*. Syracuse: Syracuse University Press.

Szasz, Thomas (1998). Summary statement and manifesto. Unpublished document dated March 1988. www.szasz.com/manifesto.html

Szasz, Thomas S. (2004). Knowing what ain't so: R. D. Laing and Thomas Szasz. *Psychoanalytic Review*, 91(3), 331–46.

Tabb, Kathryn (2015). Psychiatric progress and the assumption of diagnostic discrimination. *Philosophy of Science*, 82(5), 1047–58.

Tabb, Kathryn (2019). Philosophy of psychiatry after diagnostic kinds. *Synthese*, 196(6), 2177–99.

Tabb, Kathryn, & Schaffner, Kenneth F. (2017). Causal pathways, random walks and tortuous paths: Moving from the descriptive to the etiological. In Kenneth S. Kendler & Josef Parnas, eds., *Philosophical Issues in Psychiatry IV: Psychiatric Nosology*. Oxford: Oxford University Press, pp. 332–40.

Tekin, Şerife (2011). Self-concept through the diagnostic looking glass: Narratives and mental disorder. *Philosophical Psychology*, 24(3), 357–80.

Tekin, Şerife (2014). The missing self in Hacking's looping effects. In Harold Kincaid & Jacqueline Sullivan, eds., *Classifying Psychopathology: Mental Kinds and Natural Kinds*. Cambridge, MA: MIT Press, pp. 227–56.

Tekin, Şerife (2016). Are mental disorders natural kinds? A plea for a new approach to intervention in psychiatry. *Philosophy, Psychiatry, & Psychology*, 23(2), 147–63.

Tekin, Şerife, & Bluhm, Robin (eds.) (2019). *The Bloomsbury Companion to Philosophy of Psychiatry*. London: Bloomsbury.

Thagard, Paul (1999). *How Scientists Explain Disease*. Princeton, NJ: Princeton University Press.

Tsou, Jonathan Y. (2007). Hacking on the looping effects of psychiatric classifications: What is an interactive and indifferent kind? *International Studies in the Philosophy of Science*, 21(3), 329–44.

Tsou, Jonathan Y. (2011). The importance of history for philosophy of psychiatry: The case of the DSM and psychiatric classification. *Journal of the Philosophy of History*, 5(3), 446–70.

Tsou, Jonathan Y. (2012). Intervention, causal reasoning, and the neurobiology of mental disorders: Pharmacological drugs as experimental instruments. *Studies in History and Philosophy of Biological and Biomedical Sciences*, 43(2), 542–51.

Tsou, Jonathan Y. (2013). Depression and suicide are natural kinds: Implications for physician-assisted suicide. *International Journal of Law and Psychiatry*, 36(5-6), 461–70.

Tsou, Jonathan Y. (2015). DSM-5 and psychiatry's second revolution: Descriptive vs. theoretical approaches to psychiatric classification. In Steeves Demazeux & Patrick Singy, eds., *The DSM-5 in Perspective: Philosophical Reflections on the Psychiatric Babel*. Dordrecht: Springer, pp. 43–63.

Tsou, Jonathan Y. (2016). Natural kinds, psychiatric classification and the history of the *DSM*. *History of Psychiatry*, 27(4), 406–24.

Tsou, Jonathan Y. (2017). Pharmacological interventions and the neurobiology of mental disorders. In Ioan Opris & Manuel F. Casanova, eds., *The Physics of the Mind and Brain Disorders: Integrated Neural Circuits Supporting the Emergence of Mind*. Cham: Springer, pp. 613–28.

Tsou, Jonathan Y. (2019). Philosophy of science, psychiatric classification, and the DSM. In Şerife Tekin & Robyn Bluhm, eds., *The Bloomsbury Companion to Philosophy of Psychiatry*. London: Bloomsbury, pp. 177–96.

Tsou, Jonathan Y. (2020). Social construction, HPC kinds, and the projectability of human categories. *Philosophy of the Social Sciences*, 50(2), 115–37.

Varga, Somogy (2012). Evolutionary psychiatry and depression: Testing two hypotheses. *Medicine, Health Care and Philosophy*, 15(1), 41–52.

Wakefield, Jerome C. (1992a). Disorder as harmful dysfunction: A conceptual critique of *DSM-III-R*'s definition of mental disorder. *Psychological Review*, 99(2), 232–47.

Wakefield, Jerome C. (1992b). The concept of mental disorder: On the boundary between biological facts and social values. *American Psychologist*, 47(3), 373–88.

Wakefield, Jerome C. (1999). Evolutionary versus prototype analyses of the concept of disorder. *Journal of Abnormal Psychology*, 108(3), 374–99.

Wakefield, Jerome C. (2000). Spandrels, vestigial organs, and such: Reply to Murphy and Woolfolk's "The harmful dysfunction analysis of mental disorder." *Philosophy, Psychiatry, & Psychology*, 7(4), 253–69.

Washington, Natalia (2016). Culturally unbound: Cross-cultural cognitive diversity and the science of psychopathology. *Philosophy, Psychiatry, & Psychology*, 23(3), 165–79.

Watson, David (2009). Differentiating the mood and anxiety disorders: A quadripartite model. *Annual Review of Clinical Psychology*, 5, 221–47.

Watson, David, O'Hara, Michael, & Stuart, Scott (2008). Hierarchical structures of affect and psychopathology and their implications for the classification of emotional disorders. *Depression and Anxiety*, 25(4), 282–8.

Weinberger, Daniel R. (1987). Implications of normal brain development for schizophrenia. *Archives of General Psychiatry*, 44(7), 660–9.

Weissman, Myrna M., Bland, Roger C., Canino, Glorisa C., et al. (1996). Cross-national epidemiology of major depression and bipolar disorder. *Journal of the American Medical Association*, 276(4), 293–9.

WHO (2017). *Depression and Other Common Mental Disorders: Global Health Estimates*. Geneva: World Health Organization.

Wilson, Mark (1993). DSM-III and the transformation of American psychiatry: A history. *American Journal of Psychiatry*, 150(3), 399–410.

Wilson, Robert A. (1999). Realism, essence, and kind: Resuscitating species essentialism? In Robert A. Wilson, ed., *Species: New Interdisciplinary Essays*. Cambridge, MA: MIT Press, pp. 187–207.

Wilson, Robert A., Matthew J. Barker, and Ingo Brigandt. 2007. When traditional essentialism fails: Biological natural kinds. *Philosophical Topics*, 35(1–2), 189–215.

Wimsatt, William C. (1972). Teleology and the logical structure of function statements. *Studies in History and Philosophy of Science*, 3(1), 1–80.

Wimsatt, William C. (1981). Robustness, reliability, and overdetermination. In Marilynn B. Brewer & Barry E. Collins, eds., *Scientific Inquiry and the Social Sciences: A Volume in Honor of Donald T. Campbell*. San Francisco, CA: Jossey-Bass, pp. 124–63.

Wimsatt, William C. (2007). *Re-engineering Philosophy for Limited Beings: Piecewise Approximations to Reality*. Cambridge, MA: Harvard University Press.

Wong, David T., Perry, Kenneth W., & Bymaster, Frank P. (2005). The discovery of fluoxetine hydrochloride (Prozac). *Nature Reviews Drug Discovery*, 4(9), 764–74.

Woodward, James (2003). *Making Things Happen: A Theory of Causal Explanation*. Oxford: Oxford University Press.

Wray, N. R., Pergadia, M. L., Blackwood, D. H. R., et al. (2012). Genome-wide association study of major depressive disorder: New results, meta-analysis, and lessons learned. *Molecular Psychiatry*, 17(1), 36–48.

Zachar, Peter (2000). Psychiatric disorders are not natural kinds. *Philosophy, Psychiatry, & Psychology*, 7(3), 167–82.

Zachar, Peter (2014). *The Metaphysics of Psychopathology*. Cambridge, MA: MIT Press

Zachar, Peter, & Kendler, Kenneth S. (2017). The philosophy of nosology. *Annual Reviews of Clinical Psychology*, 13, 49–71.

Acknowledgments

I am grateful to Peter Zachar, Kathryn Tabb, Anya Plutynski, Justin Garson, Şerife Tekin, Miriam Solomon, Joseph McCaffrey, Claire Pouncey, John Sadler, Brent Kious, Christopher Boorse, Peter Schwartz, Kenneth Schaffner, Robert Wilson, Trevor Pearce, and my colleagues in the Association for the Advancement of Philosophy and Psychiatry (AAPP) for helpful discussion, comments, and suggestions. I owe special thanks to Jacob Stegenga, James Broucek, Gregory Robson, and the manuscript reviewers for diligently reading and providing constructive feedback on the entire manuscript. Rudimentary formulations of the arguments articulated in this Element were developed in my doctoral dissertation (*The Reality and Classification of Mental Disorders*, 2008) supervised by William Wimsatt, Robert Richards, and Ian Hacking at the University of Chicago. I am indebted to my supervisors for their guidance and support. I am particularly appreciative to Hacking for his extensive comments and criticism.

Cambridge Elements $\equiv$

Philosophy of Science

Jacob Stegenga
University of Cambridge

Jacob Stegenga is a Reader in the Department of History and Philosophy of Science at the University of Cambridge. He has published widely on fundamental topics in reasoning and rationality and philosophical problems in medicine and biology. Prior to joining Cambridge he taught in the United States and Canada, and he received his PhD from the University of California San Diego.

About the Series

This series of Elements in Philosophy of Science provides an extensive overview of the themes, topics and debates which constitute the philosophy of science. Distinguished specialists provide an up-to-date summary of the results of current research on their topics, as well as offering their own take on those topics and drawing original conclusions.

Cambridge Elements ☰

Philosophy of Science